SIGNAL PROCESSING FOR MULTISTATIC RADAR SYSTEMS

SIGNAL PROCESSING FOR MULTISTATIC RADAR SYSTEMS

Adaptive Waveform Selection, Optimal Geometries and Pseudolinear Tracking Algorithms

NGOC HUNG NGUYEN

KUTLUYIL DOĞANÇAY

ELSEVIER

ACADEMIC PRESS
An imprint of Elsevier

Academic Press is an imprint of Elsevier
125 London Wall, London EC2Y 5AS, United Kingdom
525 B Street, Suite 1650, San Diego, CA 92101, United States
50 Hampshire Street, 5th Floor, Cambridge, MA 02139, United States
The Boulevard, Langford Lane, Kidlington, Oxford OX5 1GB, United Kingdom

Library of Congress Cataloging-in-Publication Data
A catalog record for this book is available from the Library of Congress

British Library Cataloguing-in-Publication Data
A catalogue record for this book is available from the British Library

ISBN: 978-0-12-815314-7

For information on all Academic Press publications
visit our website at https://www.elsevier.com/books-and-journals

Publisher: Mara Conner
Acquisition Editor: Tim Pitts
Editorial Project Manager: John Leonard
Production Project Manager: Kamesh Ramajogi
Designer: Matthew Limbert

Typeset by VTeX

Contents

About the Authors

Ngoc Hung Nguyen received the B.E. degree (Hons.) in electrical and electronic engineering from the University of Adelaide, Australia, in 2012, and the Ph.D. degree in telecommunications from the University of South Australia, Australia, in 2016. He is currently a Researcher with the Defence Science and Technology Group (DST Group), Australia, and an Adjunct Research Fellow with the University of South Australia. Before joining DST Group in 2019, he was a Research Fellow and Lecturer with the University of South Australia. His research interests include statistical and adaptive signal processing, compressive sensing, and estimation theory with emphasis on target localization and tracking, sensor management, and radar imaging. Dr. Nguyen currently serves on the Editorial Board of Digital Signal Processing.

Kutluyıl Doğançay received the B.S. degree with honors in electrical and electronic engineering from Boğaziçi University, Istanbul, Turkey, in 1989, the M.Sc. degree in communications and signal processing from Imperial College, The University of London, UK, in 1992, and the Ph.D. degree in telecommunications engineering from The Australian National University, Canberra, ACT, Australia, in 1996. Since November 1999, he has been with the School of Engineering, University of South Australia, where he is a professor and discipline leader of electrical and mechatronic engineering. His research interests span statistical and adaptive signal processing with applications in defence and communication systems. Dr. Doğançay received the Best Researcher Award of School of Engineering, University of South Australia, in 2015, and Tall Poppy Science Award of the Australian Institute of Political Science in 2005. He was the Tutorials Chair of the IEEE Statistical Signal Processing Workshop (SSP 2014), and the Signal Processing and Communications Program Chair of the 2007 Information, Decision and Control Conference. He serves on the Editorial Board of Signal Processing and the EURASIP Journal on Advances in Signal Processing. From 2009–2015 he was an elected member of the Signal Processing Theory and Methods (SPTM) Technical Committee of the IEEE Signal Processing Society. He is currently an associate member of the Sensor Array and Multichannel (SAM) Technical Committee and a member of the IEEE ComSoc Signal Processing for Communications and Electronics Technical Committee. Dr. Doğançay is the EURASIP liaison for Australia.

Preface

Radar is an electromagnetic sensor for detecting and ranging objects by radiating electromagnetic waves to illuminate the scene of interest and receiving reflected echoes from the objects. Since its invention at the beginning of the 20th century, radar has played a fundamental and prominent role in a wide range of applications in both the civilian and military domains thanks to its capability to operate under all weather conditions, at day and night, and to detect, track, and image targets with high accuracy at long stand-off ranges. Multistatic radar is a particular type of radar that incorporates multiple spatially distributed transmitters and receivers. The spatial diversity provided by the separation of transmitters and receivers not only offers several advantages over conventional monostatic radar, but it also gives system designers extra degrees of freedom to optimize the radar system design for specific applications. Moreover, great progress has been made toward solving the complexity issues of multistatic radar, thus making multistatic radar systems perfectly feasible in real-world situations.

The conventional definition of the ambiguity function with respect to time delay and Doppler shift provides relatively little information about multistatic radar performance because time delay and Doppler shift in the case of multistatic radar have nonlinear relationships with target position and velocity. Instead, the multistatic ambiguity function is often expressed in terms of target position and velocity. In this form, the multistatic ambiguity function is not only dependent on the radar waveform, but it is also strongly influenced by the radar-target geometry. This emphasizes the importance of geometry for the performance of multistatic radar.

Inspired by the unique peculiarities of multistatic radar in relation to nonlinear signal processing, waveform and geometry, this book presents modern signal processing techniques for multistatic tracking radar systems with the pivotal theme of performance optimization via waveform adaptation, geometry optimization and pseudolinear tracking algorithms. The core of this book is structured in three parts. Part 1 (Adaptive waveform selection) presents several adaptive waveform selection algorithms for various multistatic tracking problems. Part 2 (Optimal geometry analysis) addresses the problem of radar-target geometry optimization for multistatic time-of-arrival based target localization. Part 3 (Pseudolinear tracking algorithms) focuses on the design and analysis of inherently-stable and low-complexity closed-form tracking and localization algorithms for passive multistatic radar systems.

Simulation examples provided within the book are accompanied by supplementary MATLAB codes which are available online at the webpage of this book on the Elsevier website https://www.elsevier.com/books-and-journals/book-companion/9780128153147. These MATLAB codes are given on an as-is basis, and no warranties are claimed. The contributors of the codes will not be held liable for any damage caused.

The authors would like to extend their sincere gratitude to Professors Fulvio Gini and Maria Sabrina Greco of the University of Pisa and Doctor Hatem Hmam of the Australian Defence Science and Technology Group for their expert peer-review and feedback on the book proposal. The authors are grateful to the Editorial Project Manager of Academic Press, John Leonard, and all of the publishing team for their support in the publication of this book. In addition, the first author expresses his loving gratitude and dedicates this book to his wife, Lien, for her love and encouragement. The second author acknowledges the support and understanding of his family during the writing of this book.

List of Abbreviations and Symbols

Abbreviations

2D	Two-dimensional
3D	Three-dimensional
AOA	Angle-of-arrival
BCPLE	Bias-compensated PLE
CRLB	Cramér-Rao lower bound
DEKF	Diffusion extended Kalman filter
EKF	Extended Kalman filter
FDOA	Frequency-difference-of-arrival
FIM	Fisher information matrix
IMM	Interacting multiple model
IV	Instrumental variable
IVE	IV estimator
LFM	Linear frequency-modulated
LOS	Line of sight
MIMO	Multiple-input multiple-output
MLE	Maximum likelihood estimator
MSE	Mean squared error
PDA	Probabilistic data association
PLE	Pseudolinear estimator
RMSE	Root mean squared error
Rx	Receiver
TB	Time-bandwidth
TDOA	Time-difference-of-arrival
TLS	Total least-squares
TMA	Target motion analysis
TOA	Time-of-arrival
Tx	Transmitter
UAV	Unmanned aerial vehicle
WIVE	Weighted instrumental variable estimator

Symbols

$\mathbb{R}^n$	Set of real vectors of dimension n
x	A scalar
$\boldsymbol{x}$	A vector
$\boldsymbol{X}$	A matrix
$\boldsymbol{I}_{N \times N}$ or $\boldsymbol{I}$	An identity matrix of dimension $N \times N$ or of appropriate dimension
$\mathbf{0}_{N \times M}$ or $\mathbf{0}$	Zero matrix of dimension of $N \times M$ or of appropriate dimension
$\mathbf{1}_{N \times M}$ or $\mathbf{1}$	A matrix of ones of dimension $N \times M$ or of appropriate dimension
$\boldsymbol{x}(i)$	The ith element of vector $\boldsymbol{x}$

$\boldsymbol{X}(i)$	The ith diagonal element of matrix $\boldsymbol{X}$
$\boldsymbol{X}(i,j)$ or $\boldsymbol{X}^{[i,j]}$	The (i,j)th element of matrix $\boldsymbol{X}$
$\boldsymbol{x}(i:j)$	A sub-vector formed by the ith to jth elements of vector $\boldsymbol{x}$
$\boldsymbol{X}(i,:)$	The ith row of matrix $\boldsymbol{X}$
$\mathrm{diag}(x, y, \ldots)$	A diagonal matrix with diagonal entries $x, y, \ldots$
$\mathrm{diag}(\boldsymbol{X}, \boldsymbol{Y}, \ldots)$	A block diagonal matrix with diagonal blocks $\boldsymbol{X}, \boldsymbol{Y}, \ldots$
$\mathrm{diag}(\boldsymbol{x})$	A diagonal matrix formed by the elements of vector $\boldsymbol{x}$
$\tilde{x}$	A noisy measurement of variable x
$\tilde{\boldsymbol{x}}$	A noisy measurement of vector variable $\boldsymbol{x}$
$\hat{x}$	An estimate of variable $\hat{x}$
$\hat{\boldsymbol{x}}$	An estimate of vector variable $\hat{\boldsymbol{x}}$
$\boldsymbol{X}^{\circ}$ or $\boldsymbol{X}_{\circ}$	The noise-free version of matrix $\boldsymbol{X}$
$\lVert \cdot \rVert$	Euclidean norm
$(\cdot)^T$	Matrix transpose
$\lvert \cdot \rvert$	Matrix determinant
$\mathrm{trace}(\cdot)$	Matrix trace
$\log(\cdot)$	Base-10 logarithm
$\ln(\cdot)$	Natural logarithm
$\exp(\cdot)$	Exponential function
$\mathrm{sgn}(\cdot)$	Signum function
$\odot$	Element-by-element (Schur) product
$E\{\cdot\}$	Statistical expectation

CHAPTER 1

Introduction

Contents

1.1 Historical background

Over its history of more than a century, radar has played a fundamental and prominent role in a wide range of civilian and military applications thanks to its capability to operate in all weather, day and night, and to detect, track, and image targets with high accuracy at long stand-off ranges. Nowadays, the use of radar can be found in weather forecasting, remote sensing and mapping, astronomy, aerial and terrestrial traffic control, air defense and weapon control, automotive sensor systems, high-resolution target imaging and recognition, airborne collision avoidance systems, and eldercare and assisted living, to name but a few [1–12].

Radar is essentially an electromagnetic sensor for detecting and ranging objects by radiating electromagnetic wave to illuminate the scene of interest and receiving reflected echoes from the objects. By processing the received signal, radar can make a decision on whether one or more targets are present in the scene of interest, and determine the position and velocity and possibly the size, shape and features of the detected targets. In terms of system geometry, radar can be categorized into three different types: monostatic radar, bistatic radar and multistatic radar. Monostatic radar has a transmitter and a receiver which are collocated at the same site. In contrast, bistatic radar is a radar that operates with a transmitter and a receiver located separately at different sites. Multistatic radar is a generalization of bistatic radar by incorporating multiple transmitters and receivers at different locations while having overlapping spatial coverage, as illustrated in Fig. 1.1.

By conventional definition, multistatic radar can be viewed as a system of several individual transmitter–receiver pairs that operate independently [13,14]. Detection, target estimation, and other high-level target information from these transmitter–receiver pairs are communicated to a central processor, where they are combined to improve detection and

Signal Processing for Multistatic Radar Systems
https://doi.org/10.1016/B978-0-12-815314-7.00008-1

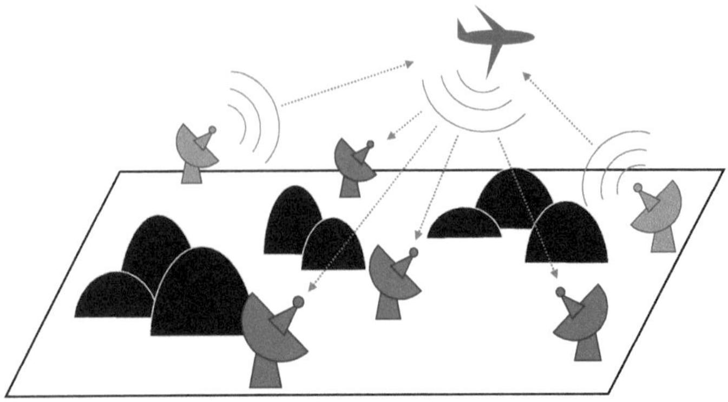

Figure 1.1 An illustration of a multistatic radar system.

estimation performance. This type of data fusion is often considered as non-coherent processing. For example, multistatic radar can use multilateration to estimate the target kinematic state (i.e., position and velocity) based on range, Doppler and/or angle measurements obtained independently at different transmitter–receiver pairs within the system. MIMO (multiple-input multiple-output) radar with widely separated antennas is another type of multistatic radar with a different design that distinguishes it from multistatic radar by joint processing at the signal level for both transmission and reception, as well as its close connection to MIMO communications [15,16]. However, this type of multistatic radar is not covered in this book.

The spatial diversity provided by the separation of transmitters and receivers not only brings a number of advantages to bistatic and multistatic radars over monostatic radar, but also gives system designers extra degrees of freedom to optimize the radar system design for specific applications [13, 14]. Since the receivers are not collocated with the transmitters, bistatic and multistatic radars are able to counter retrodirective jammers which sense the radar signals from the transmitters and direct jamming signals back towards the transmitters. The bistatic/multistatic geometry also enhances the radar cross section for stealthy targets, thus improving the detection performance. Moreover, the target kinematic state can be estimated with a high accuracy by multistatic radar thanks to the exploitation of multilateration from measurements collected by multiple transmitter–receiver pairs, each having a different bistatic geometry with respect to the target. In addition, multistatic radar can avoid the unfavorable geometry of bistatic radar in which the target is located near the line-of-sight between the transmitter and receiver.

Bistatic and multistatic radars can incorporate their own dedicated transmitters or exploit illuminators of opportunity from other transmission sources [13,14,17]. The illuminators of opportunity may come from existing radar systems (either cooperative or non-cooperative) or from commercial non-cooperative broadcast and communications signals. The radar systems that utilize illuminators of opportunity are commonly known as passive. In a passive configuration, bistatic and multistatic radar can operate in civilian areas, particularly in large cities with multiple airports, where radar transmission may not be allowed. In addition, exploiting the illuminators of opportunity from other existing transmission sources eliminates the construction and operation costs for transmitters and associated equipment.

These advantages of bistatic and multistatic radar come at the expense of increased complexity for system hardware and signal processing, particularly in terms of direct-path signal cancellation and transmitter–receiver synchronization. However, great progress has been made towards solving these complexity issues thanks to the tremendous amount of research dedicated to the topics of beamforming, integrated circuit design, and the Global Positioning System, thus making bistatic and multistatic radar systems perfectly feasible in real-world situations [13].

Bistatic and multistatic radars have experienced periodic resurgences throughout the history of radar. The very first radars were of the bistatic type. Prior to and during World War II, several continuous-wave bistatic radars were developed and deployed in many countries including the United States, United Kingdom, Soviet Union, France, Germany, Japan and Italy. Since the invention of radar duplexer, which allows the use of pulsed waveforms with a single common antenna for both transmission and reception, monostatic radar has dominated the field of radar research mainly due to the advantages of single-site operation. Consequently, all bistatic radar work was discontinued by the end of World War II. The first resurgence of bistatic and multistatic radar started in the 1950s with applications in air defense and ballistic missile launch warning, tactical semiactive homing missiles, and test range instrumentation and satellite tracking. The second resurgence happened in the 1970s and 1980s. A number of experimental bistatic radar systems were tested (but not deployed) in response to the retro-jamming and antiradiation missile threats. The Multistatic Measurement System for ballistic missile tracking, which is a multistatic radar hitchhiker, was deployed in the 1980s. The first experiment of using broadcast transmitters as illuminators of opportunity was conducted during this resurgence period. Since the beginning of the third resurgence in the mid-1990s, extensive research studies have been devoted to bistatic and multistatic radars including image focusing and motion compensation techniques for bistatic Synthetic Aperture Radar, adaptive clutter cancellation methods for bistatic moving target indication, and exploitation of commercial broadcast transmitters for passive radar. A comprehensive review of the history of radar, including bistatic and multistatic radars, is available in [13, 14,17].

1.2 Purpose and scope

The ambiguity function is an important and indispensable tool for radar performance analysis [8,18,19]. The ambiguity function essentially characterizes the auto-correlation of the complex envelope of the radar waveform with its shifted copy in time and frequency. The point target response of the waveform is expressed by the ambiguity function as a two-dimensional function of time delay and Doppler shift (or equivalently target range and radial velocity). Based on the shape of the ambiguity function, radar performance can be evaluated in terms of estimation accuracy, target resolution, and clutter cancellation. In monostatic radar, where time delay and Doppler shift are proportional to target range and radial velocity, respectively, the ambiguity function is only dependent on the radar waveform without having any influences from the radar-target geometry. In terms of time delay and Doppler shift, the ambiguity function for bistatic and multistatic radars remains identical to that of monostatic radar. However, this definition of ambiguity function provides relatively little information about radar performance because time delay and Doppler shift have nonlinear relationships with target position and velocity parameters [20,21]. Depending on the application, different coordinate systems can be used to characterize target position and velocity, for example, the North-referenced system, the Cartesian system, and spherical coordinates. Therefore, for bistatic and multistatic radars, the ambiguity functions are often expressed in terms of target position and velocity in these coordinate systems. In such forms, the bistatic and multistatic ambiguity functions are not only dependent on the radar waveform, but also strongly influenced by the radar-target geometry [20, 21]. This highlights the importance of geometry to the performance of multistatic radar.

The focus of this book is on modern signal processing techniques for multistatic tracking radar systems. Inspired by the unique peculiarities of multistatic radar as highlighted above, the pivotal theme of the topics presented in this book is performance optimization via waveform adaptation, geometry optimization and pseudolinear tracking algorithms. The core of this book is structured into three parts:

- **Part 1: Adaptive waveform selection** (Chapters 2–5) presents a number of adaptive waveform selection algorithms for various multistatic tracking scenarios including maneuvering target tracking in Chapter 2, target tracking in clutter in Chapter 3, target tracking with Cartesian estimates in Chapter 4, and distributed target tracking in Chapter 5. The objective is to optimize tracking performance via minimizing the mean squared error (MSE) of the target kinematic state estimate in the Cartesian coordinates. This is achieved by minimizing the trace of the target state error covariance matrix which can be computed from the update equations of the tracking algorithms using the Cramér–Rao lower bound (CRLB) of radar measurement errors.
- **Part 2: Optimal geometry analysis** (Chapters 6–7) addresses the problem of radar-target geometry optimization for multistatic target localization. The optimal geometries for time-of-arrival (TOA)-based localization are analytically derived for two multistatic configurations: (i) a single transmitter and multiple receivers in Chapter 6, and (ii) multiple independent bistatic channels in Chapter 7. The optimal geometry analyses are conducted based on minimizing the area of estimation confidence region, which is equivalent to maximizing the determinant of the Fisher information matrix (FIM).
- **Part 3: Pseudolinear tracking algorithms** (Chapters 8–9) focuses on the design and analysis of inherently-stable and low-complexity closed-form tracking and localization algorithms for passive multistatic radar systems. Closed-form batch estimators are presented in Chapter 8 for multistatic target motion analysis of a constant-velocity target utilizing angle-of-arrival (AOA), time-difference-of-arrival (TDOA) and frequency-difference-of-arrival (FDOA) measurements under the framework of pseudolinear estimation and instrumental-variable estimation. The method of closed-form least-squares estimation and its bias-reduced variants are covered in Chapter 9 for multistatic target localization with TDOA measurements.

We assume that the reader has a background in digital signal processing, linear and matrix algebra, and probability and random processes, as well as a basic understanding of radar system and signal processing (e.g., [1,2]), and estimation and tracking (e.g., [22–24]).

1.3 Outline of book

A brief overview and summary of the remaining chapters of the book is given below.

Part 1: Adaptive waveform selection

Chapter 2: Waveform selection for multistatic tracking of a maneuvering target

This chapter considers the problem of adaptive waveform selection for multistatic tracking of a maneuvering target. An interacting multiple model algorithm incorporating extended Kalman filters is employed to deal with the nonlinearity between the radar observation vector and the target state vector and to handle multiple models of target dynamics. Three waveform selection schemes are presented based on minimizing the traces of the most-likely, minimax and combined covariance matrices of the interacting multiple model algorithm. Monte Carlo simulation examples are presented to demonstrate the performance advantages of the waveform selection schemes over conventional fixed waveforms.

Chapter 3: Waveform selection for multistatic target tracking in clutter

This chapter presents an optimal waveform selection algorithm for multistatic target tracking in a cluttered environment. Due to the false-alarm measurements caused by clutter, the exact track error covariance matrix is not available during the waveform selection process (i.e., prior to the actual waveform transmission). Therefore, an expected track error covariance matrix computed via the modified Riccati equation is exploited for waveform selection. Simulation examples corroborate the effectiveness of the presented waveform selection algorithm.

Chapter 4: Waveform selection for multistatic target tracking with Cartesian estimates

This chapter is concerned with multistatic tracking using Cartesian estimates, where only desired measurements from two receivers in the system are selected and utilized to compute a Cartesian state estimate for target dynamics which is then processed by a linear Kalman tracker. Joint adaptive selection of radar waveform and Cartesian estimate is presented to optimize tracking performance. The CRLBs for Cartesian estimates are derived and exploited in the development of the joint adaptive selection scheme. The tracking system under consideration provides three important advantages; viz., (i) minimal requirement for bandwidth and power consumption in transmitter–receiver communication links by utilizing only one Cartesian estimate for target tracking, (ii) performance optimization brought about by the joint adaptive selection of radar waveform and Cartesian estimate, and (iii) inherent benefits of linear estimation, e.g., stability, from the use of linear Kalman filter. The performance advantages of the presented tracking system are demonstrated via simulation examples.

Chapter 5: Waveform selection for distributed multistatic target tracking

This chapter addresses the waveform selection problem in the context of distributed multistatic target tracking, where the receivers form a connected network in which they can communicate their radar measurements and track estimates to their immediate neighbors. A method of performing joint target tracking and waveform selection over the receivers in a fully decentralized manner is presented. Specifically, target tracking and waveform selection are performed at each receiver exploiting the information data available in its closed neighborhood. Such a distributed processing strategy offers many important advantages in terms of cost savings for the operation and maintenance of the system, as well as increasing the robustness of the system to link or node failures. Simulation examples are given to corroborate the performance of the presented method.

Part 2: Optimal geometry analysis

Chapter 6: Optimal geometries for multistatic target localization with one transmitter and multiple receivers

This chapter analyzes the optimal geometries for the problem of target localization using TOA measurements collected from a multistatic radar system with one transmitter and multiple receivers. The optimal geometry analysis is conducted based on maximizing the determinant of the FIM which effectively minimizes the area of the estimation confidence region. Rotating the coordinate system so that the bearing angle of the transmitter relative to the target is zero, an optimal geometry is formed by placing some receivers collinearly with the target at the bearing angle of 60 degrees and the remaining receivers at −60 degrees. The accuracy of the analytical findings is confirmed via simulation studies including numerical solutions and sensor trajectory optimization.

Chapter 7: Optimal geometries for multistatic target localization by independent bistatic channels

This chapter derives the optimal geometries for the target localization problem using TOA measurements collected from a multistatic radar system consisting of multiple independent bistatic channels. The optimal angular geometries for this multistatic configuration are obtained when the target is collinear with the transmitter and receiver of each bistatic channel and the target is located at either end. The problem reduces to optimizing angular separation between different bistatic channels which is readily solved by referring to the results of optimal angular separation for AOA localization. Simulation examples on sensor trajectory optimization are presented to verify the analytical findings.

Part 3: Pseudolinear tracking algorithms

Chapter 8: Batch track estimators for multistatic target motion analysis

This chapter considers the problem of target motion analysis for a constant-velocity target by a passive multistatic radar system using AOA, TDOA and FDOA measurements. Four batch estimators are presented including the pseudolinear estimator (PLE), bias-compensated PLE (BCPLE), weighted instrumental-variable estimator (WIVE) and maximum likelihood estimator (MLE). Compared to the iterative MLE, the PLE, BCPLE and WIVE are closed-form with inherent stability and computation advantages. The PLE suffers from severe bias problems due to the correlation between the measurement matrix and pseudolinear noise vector. The BCPLE alleviates the PLE bias by estimating and subtracting the PLE instantaneous bias. The BCPLE estimate is incorporated into the WIVE to produce asymptotically unbiased estimates of target motion parameters. The WIVE is analytically shown to be asymptotically efficient under small measurement noise. The performance of the estimators is evaluated via numerical simulation examples. The WIVE produces the best performance among the closed-form estimators, while exhibiting performance on a par with the computationally more demanding MLE.

Chapter 9: Closed-form solutions for multistatic target localization with time-difference-of-arrival measurements

This chapter focuses on closed-form least-squares estimation techniques for multistatic target localization using TDOA measurements. To overcome the computational burden and instability of the MLE, the nonlinear TDOA measurement equations are algebraically rearranged into a set of linear equations by introducing a nuisance parameter, thus enabling linear least-squares to be performed in closed-form. Such a transformation of the TDOA measurement equations creates two technical challenges. First, the nuisance parameter is dependent on the true target position and this dependence must be taken into account in order to obtain efficient estimates. Second, the measurement matrix in the linearized equations is correlated with the pseudolinear noise vector, leading to bias problems in the linear least-squares solution. In this chapter, we cover a range of algorithms that can effectively tackle these two challenges. Simulation examples are presented to corroborate the performance benefits of the algorithms.

PART 1

Adaptive waveform selection

Background and purpose

Waveform diversity is the concept of dynamically adapting the radar waveform to optimize the radar performance and confront the nonstationarity and uncertainty of the environment [8,18]. Waveform diversity has been exploited across different aspects of radar signal processing from detection, tracking to classification (see e.g., [8,18] and the references therein). Various waveform parameters can be used for waveform diversity, including pulse repetition frequency, carrier frequency, bandwidth, pulse width, amplitude, coding, spatial characteristics, polarization, jitter, and frequency shaping. While waveform adaptation and agility have been only recently applied to synthetic sensing systems like radar and sonar (which have less than 100 years of history), waveform diversity in fact has been being exploited in nature by echolocating mammals like whales, dolphins and bats for millions of years.

In target tracking, the tracker incorporates information about the target (e.g., range, Doppler and/or angle information) in a sequential manner, while the radar-target geometry and sensing environment are constantly changing due to radar and/or target motion. Therefore, dynamic adaptation of waveforms is especially advantageous in target tracking applications as it allows the transmitted waveform to be adjusted on a pulse-to-pulse basis to accommodate changes in tracker requirements [25–32]. Waveform adaptation in target tracking is motivated by the fact that different waveforms have different resolution properties and thus induce different errors in radar measurements. The idea is to design or select the waveform to ensure small measurement errors in the dimensions of the target kinematic state where the uncertainty of the tracker is large while higher levels of errors may be tolerated for other dimensions with less uncertainty. Dynamic waveform adaptation is commonly considered under the framework of control theory, where the choice of the waveform to be transmitted in the next time step is made by optimizing a waveform-dependent cost function such as the mean squared error of the target state estimation error [28–34]. This formulation of the problem results in a feedback loop where the waveform selected at the end of the current time step affects the radar and tracker performance in the next time step which in turn influences the next waveform selection. For the simple case of one-dimensional target tracking with a linear observation model, closed-form solutions are available [28]. However, brute force optimization techniques are often required in most modern tracking scenarios because of the nonlinearity of observation models, the presence of clutter or the imperfection of detection [29–32]. For this reason, the problem is commonly referred to as adaptive waveform selection. The objective is to select the best waveform candidate within a library of available waveforms which can optimize the tracking performance at the next time step. A waveform library may consist of different waveform classes, or simply a collection of waveforms of the same type with different waveform parameters.

In this part of the book, we present several adaptive waveform selection algorithms for different multistatic tracking scenarios based on the control theoretical approach:

- Chapter 2: maneuvering target tracking,
- Chapter 3: target tracking in clutter,
- Chapter 4: target tracking with Cartesian target state estimates,
- Chapter 5: target tracking in a distributed multistatic system.

The objective is to minimize the MSE of the target kinematic state estimate in the Cartesian coordinates by minimizing the trace of the target state error covariance matrix which can be computed from the update equations of the tracking algorithms using the CRLB of radar measurement errors.

CHAPTER 2

Waveform selection for multistatic tracking of a maneuvering target

Contents

2.1 Introduction and system overview

This chapter considers the problem of adaptive waveform selection for maneuvering target tracking with a multistatic radar system consisting of a single transmitter (Tx) and multiple receivers (Rx). The overall system configuration is depicted in Fig. 2.1. At each receiver, time delay, Doppler shift and angle-of-arrival of the received signal (which is reflected back from the target) are estimated by an antenna array. The receivers then communicate their measurements to a central processor located at the transmitter site to perform target tracking and waveform selection. To track a maneuvering target, the interacting multiple model (IMM) algorithm [23,24,35] is employed. After that, waveform selection is performed to determine the next waveform to be transmitted in order to minimize the MSE of the target state estimate.

For the sake of simplicity, we consider a two-dimensional (2D) tracking scenario with the radar-target geometry depicted in Fig. 2.2. It is straightforward to extend the problem to three-dimensional (3D) scenarios. In Fig. 2.2, $\boldsymbol{p}_k = [p_{x,k}, p_{y,k}]^T$ and $\boldsymbol{v}_k = [v_{x,k}, v_{y,k}]^T$ denote the unknown target position and velocity at discrete-time instant $k \in \{0, 1, 2, \dots\}$, respectively.

Signal Processing for Multistatic Radar Systems
https://doi.org/10.1016/B978-0-12-815314-7.00010-X

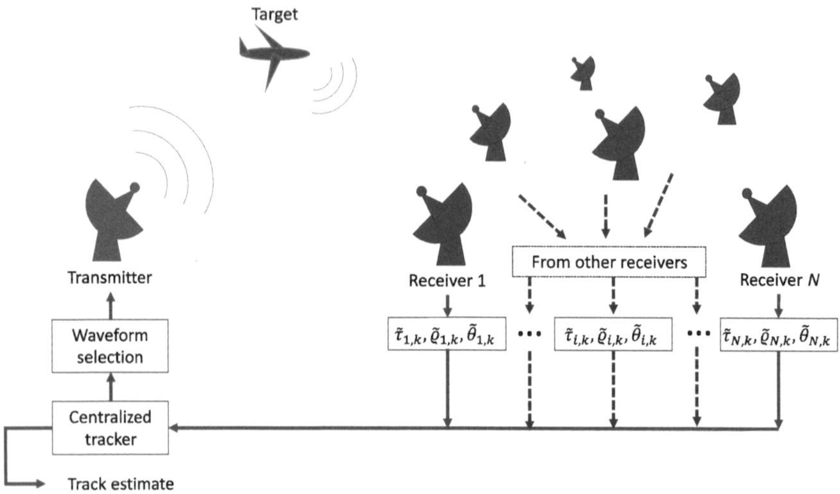

Figure 2.1 System configuration of multistatic tracking radar under consideration.

The transmitter position is denoted by $\boldsymbol{t} = [t_x, t_y]^T$, and the receiver positions by $\boldsymbol{r}_i = [r_{x,i}, r_{y,i}]^T$ $(i = 1, \ldots, N)$ where i represents the receiver index.

2.2 Bistatic radar measurements

Each receiver cooperates with the transmitter to form a bistatic channel. The time delay, Doppler shift and angle-of-arrival measurements obtained by receiver i at time instant k are given by

$$\tilde{\tau}_{i,k} = \tau_{i,k} + e_{\tau,i,k}, \tag{2.1a}$$

$$\tilde{\varrho}_{i,k} = \varrho_{i,k} + e_{\varrho,i,k}, \tag{2.1b}$$

$$\tilde{\theta}_{i,k} = \theta_{i,k} + e_{\theta,i,k}, \tag{2.1c}$$

where

$$\tau_{i,k} = \frac{1}{c}\left(\|\boldsymbol{p}_k - \boldsymbol{t}\| + \|\boldsymbol{p}_k - \boldsymbol{r}_i\|\right), \tag{2.2a}$$

$$\varrho_{i,k} = \frac{f_\circ}{c}\left(\frac{(\boldsymbol{p}_k - \boldsymbol{t})^T \boldsymbol{v}_k}{\|\boldsymbol{p}_k - \boldsymbol{t}\|} + \frac{(\boldsymbol{p}_k - \boldsymbol{r}_i)^T \boldsymbol{v}_k}{\|\boldsymbol{p}_k - \boldsymbol{r}_i\|}\right), \tag{2.2b}$$

$$\theta_{i,k} = \tan^{-1}\left(\frac{p_{y,k} - r_{y,i}}{p_{x,k} - r_{x,i}}\right). \tag{2.2c}$$

Here, $f_\circ$ is the transmission frequency, c is the speed of signal propagation, $\tan^{-1}(\cdot)$ denotes the 4-quadrant arctangent, and $\|\cdot\|$ is the Euclidean norm.

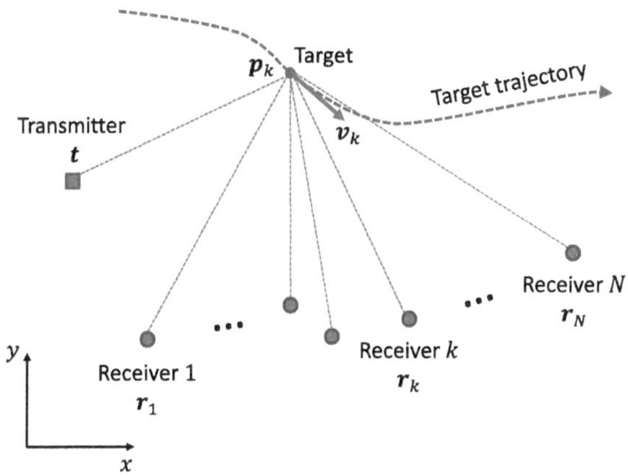

Figure 2.2 Target tracking geometry in 2D.

In (2.1), $e_{\tau,i,k}$, $e_{\varrho,i,k}$ and $e_{\theta,i,k}$ are measurement errors modeled as zero-mean Gaussian random variables.

Since the radar measurements can achieve their CRLBs approximately under mild noise conditions, we follow [28–31,36] to assume that the variances of radar measurement errors $e_{\tau,i,k}$, $e_{\varrho,i,k}$ and $e_{\theta,i,k}$ can be approximated by their corresponding CRLBs. It is important to emphasize that the radar measurement errors and CRLBs are known to be dependent on the radar transmitted waveform [19]. This dependence is discussed further in the next section, where the CRLBs for time delay and Doppler shift are explicitly expressed as functions of transmitted waveform parameters.

2.3 Bistatic ambiguity function and Cramér–Rao lower bounds

The ambiguity function is defined as the absolute value of the correlation between the complex envelope $u(t)$ of the radar transmitted waveform and its time-frequency shifted copy [19], i.e.,

$$\Theta(\tau_H, \varrho_H) = \left| \int_{-\infty}^{\infty} u(t - \tau_A) u^*(t - \tau_H) e^{-j2\pi(\varrho_H - \varrho_A)t} dt \right|. \tag{2.3}$$

Here, τ_A and ϱ_A denote the actual time delay and Doppler shift, respectively, while τ_H and ϱ_H denote the hypothesized time delay and Doppler shift, respectively. The ambiguity function $\Theta(\tau_H, \varrho_H)$ is maximized when $\tau_H = \tau_A$ and $\varrho_H = \varrho_A$. Here, we assume that the narrowband condition is satisfied, i.e., the time-bandwidth product of the transmitted signal is much smaller than the signal propagation speed divided by the time derivative of the total bistatic range to the target. The concept of ambiguity function was originally developed for monostatic radar. However, in terms of measuring time delay and Doppler shift, this ambiguity function is also applied to bistatic radar [8,37–39]. Since time delay and Doppler shift are used directly as measurement inputs to the tracker, we can characterize the accuracy of these measurements based on the ambiguity function $\Theta(\tau_H, \varrho_H)$ in (2.3).

Given a sufficiently high SNR, the Fisher information matrix $\mathbf{\Phi}_{\{\tau,\varrho\}}$, i.e., the inverse of the CRLB matrix $\boldsymbol{C}_{\{\tau,\varrho\}}$, for time delay and Doppler shift measurements is a function of both the SNR and the second derivatives of the ambiguity function $\Theta(\tau_H, \varrho_H)$ [19]:

$$\mathbf{\Phi}_{\{\tau,\varrho\}} = \boldsymbol{C}_{\{\tau,\varrho\}}^{-1} = -2\eta \left. \begin{bmatrix} \dfrac{\partial^2 \Theta(\tau_H, \varrho_H)}{\partial \tau_H^2} & \dfrac{\partial^2 \Theta(\tau_H, \varrho_H)}{\partial \tau_H \partial \varrho_H} \\ \dfrac{\partial^2 \Theta(\tau_H, \varrho_H)}{\partial \varrho_H \partial \tau_H} & \dfrac{\partial^2 \Theta(\tau_H, \varrho_H)}{\partial \varrho_H^2} \end{bmatrix} \right|_{\tau_H=\tau_A,\, \varrho_H=\varrho_A} \tag{2.4}$$

where η denotes the SNR at the receiver. Note that, consistent with the fact that the monostatic and bistatic ambiguity functions have the same form with respect to time delay and Doppler shift, the Fisher information matrix $\mathbf{\Phi}_{\{\tau,\varrho\}}$ and the CRLB matrix $\boldsymbol{C}_{\{\tau,\varrho\}}$ for time delay and Doppler shift given in (2.4) hold for both monostatic and bistatic cases.

Since the sharpness of the ambiguity function is determined by the transmitted waveform as seen in (2.3), the Fisher information matrix $\mathbf{\Phi}_{\{\tau,\varrho\}}$ and thus the CRLB matrix $\boldsymbol{C}_{\{\tau,\varrho\}}$ are also dependent on the transmitted waveform. Therefore, the CRLB matrix $\boldsymbol{C}_{\{\tau,\varrho\}}$ can be written explicitly as a function of the SNR at the receiver as well as the transmitted waveform parameters $\boldsymbol{\psi}$:

$$\boldsymbol{C}_{\{\tau,\varrho\}}(\eta, \boldsymbol{\psi}) = \mathbf{\Phi}_{\{\tau,\varrho\}}^{-1}(\eta, \boldsymbol{\psi}). \tag{2.5}$$

On the other hand, the CRLB for angle-of-arrival measurement obtained by an antenna array is independent of the transmitted waveform [40, 41], and it is inversely proportional to the SNR at the receiver, i.e.,

$$\boldsymbol{C}_{\theta}(\eta) = \frac{\sigma_\theta^2}{\eta}. \tag{2.6}$$

Although these results were originally derived for monostatic radar, they also hold for bistatic radar because the operation of bistatic radar is basically the same as that of monostatic radar in measuring time delay, Doppler shift and angle-of-arrival.

Since the CRLB for angle-of-arrival measurement is independent of the CRLBs for time delay and Doppler shift measurements, the overall CRLB matrix for time delay, Doppler shift and angle-of-arrival measurements obtained at receiver i at time instant k is given by

$$\boldsymbol{C}_{i,k}(\eta_{i,k}, \boldsymbol{\psi}_k) = \text{diag}(\boldsymbol{C}_{\{\tau,\varrho\}}(\eta_{i,k}, \boldsymbol{\psi}_k), \boldsymbol{C}_\theta(\eta_{i,k})) \tag{2.7}$$

where $\eta_{i,k}$ is the SNR at receiver i at time instant k, and $\boldsymbol{\psi}_k$ is a vector consisting of the transmitted waveform parameters at time instant k.

2.4 Target tracking

The target tracking process is performed in a centralized manner at the transmitter site after receiving radar measurements from all receivers via communication links.

2.4.1 Target dynamic model

The target motion is assumed to be governed by the kinematic model

$$\boldsymbol{x}_{k+1} = \boldsymbol{F}\boldsymbol{x}_k + \boldsymbol{w}_k \tag{2.8}$$

where $\boldsymbol{x}_k$ is the target kinematic state vector at time instant k, $\boldsymbol{F}$ is the state transition matrix and $\boldsymbol{w}_k \sim \mathcal{N}(\boldsymbol{0}, \boldsymbol{Q})$ is the process noise modeled as independent zero-mean Gaussian random variable. The target kinematic state vector may include the target position, velocity, acceleration and/or other target motion parameters, while the expressions for $\boldsymbol{F}$ and $\boldsymbol{Q}$ are dependent on the target dynamics. In what follows, we summarize three widely used kinematic models in the literature including the nearly constant velocity model, nearly constant acceleration model and nearly coordinated turn model [23,24,42].

Nearly constant velocity model

The state vector is defined by $\boldsymbol{x}_k = [\boldsymbol{p}_k^T, \boldsymbol{v}_k^T]^T$. The transition matrix $\boldsymbol{F}$ is

$$\boldsymbol{F} = \begin{bmatrix} \boldsymbol{I}_{2\times2} & T\boldsymbol{I}_{2\times2} \\ \boldsymbol{0}_{2\times2} & \boldsymbol{I}_{2\times2} \end{bmatrix} \tag{2.9}$$

and the process noise covariance matrix $\mathbf{Q}$ is

$$\mathbf{Q}=\begin{bmatrix} \frac{T^3}{3}\mathbf{Q}_o & \frac{T^2}{2}\mathbf{Q}_o \\ \frac{T^2}{2}\mathbf{Q}_o & T\mathbf{Q}_o \end{bmatrix} \tag{2.10}$$

where $\boldsymbol{I}$ and $\mathbf{0}$ are identity and zero matrices, respectively, with their dimensions specified in the superscript, and T is the sampling interval. Here, $\mathbf{Q}_o = \mathrm{diag}(q_x, q_y)$, with q_x and q_y being the power spectral densities of the process noise in x- and y-coordinates, corresponds to the level of the unknown target maneuver.

Nearly constant acceleration model

The state vector is given by $\boldsymbol{x}_k = [\boldsymbol{p}_k^T, \boldsymbol{v}_k^T, \boldsymbol{a}_k^T]^T$ where $\boldsymbol{a}_k = [a_{x,k}, a_{y,k}]^T$ denotes the acceleration of the target at time instant k. The expressions of $\boldsymbol{F}$ and $\mathbf{Q}$ for this model are

$$\boldsymbol{F}=\begin{bmatrix} \boldsymbol{I}_{2\times2} & T\boldsymbol{I}_{2\times2} & \frac{T^2}{2}\boldsymbol{I}_{2\times2} \\ \mathbf{0}_{2\times2} & \boldsymbol{I}_{2\times2} & T\boldsymbol{I}_{2\times2} \\ \mathbf{0}_{2\times2} & \mathbf{0}_{2\times2} & \boldsymbol{I}_{2\times2} \end{bmatrix} \tag{2.11}$$

and

$$\mathbf{Q}=\begin{bmatrix} \frac{T^5}{20}\mathbf{Q}_o & \frac{T^4}{8}\mathbf{Q}_o & \frac{T^3}{6}\mathbf{Q}_o \\ \frac{T^4}{8}\mathbf{Q}_o & \frac{T^3}{3}\mathbf{Q}_o & \frac{T^2}{2}\mathbf{Q}_o \\ \frac{T^3}{6}\mathbf{Q}_o & \frac{T^2}{2}\mathbf{Q}_o & T\mathbf{Q}_o \end{bmatrix}. \tag{2.12}$$

Nearly coordinated turn model

The state vector is $\boldsymbol{x}_k = [\boldsymbol{p}_k^T, \boldsymbol{v}_k^T]^T$. The transition matrix $\boldsymbol{F}$ is given by

$$\boldsymbol{F}=\begin{bmatrix} 1 & 0 & \frac{\sin(\omega T)}{\omega} & -\frac{1-\cos(\omega T)}{\omega} \\ 0 & 1 & \frac{1-\cos(\omega T)}{\omega} & \frac{\sin(\omega T)}{\omega} \\ 0 & 0 & \cos(\omega T) & -\sin(\omega T) \\ 0 & 0 & \sin(\omega T) & \cos(\omega T) \end{bmatrix} \tag{2.13}$$

and the process noise covariance matrix $\mathbf{Q}$ given by

$$\mathbf{Q} = q \begin{bmatrix} Q_{11} & Q_{12} & Q_{13} & Q_{14} \\ Q_{21} & Q_{22} & Q_{23} & Q_{24} \\ Q_{31} & Q_{32} & Q_{33} & Q_{34} \\ Q_{41} & Q_{42} & Q_{43} & Q_{44} \end{bmatrix} \tag{2.14}$$

with

$$Q_{11} = Q_{22} = \frac{2(\omega T - \sin(\omega T))}{\omega^3} \tag{2.15a}$$

$$Q_{12} = Q_{21} = Q_{34} = Q_{43} = 0 \tag{2.15b}$$

$$Q_{13} = Q_{24} = Q_{31} = Q_{42} = \frac{1 - \cos(\omega T)}{\omega^2} \tag{2.15c}$$

$$Q_{14} = Q_{41} = \frac{\omega T - \sin(\omega T)}{\omega^2} \tag{2.15d}$$

$$Q_{23} = Q_{32} = -\frac{\omega T - \sin(\omega T)}{\omega^2} \tag{2.15e}$$

$$Q_{33} = Q_{44} = T. \tag{2.15f}$$

In the above model, ω is the angular turn rate and is presumed to be a known constant. For the case that the turn rate ω is not a known constant, the coordinated turn model becomes nonlinear where the turn rate ω is incorporated as an element of an augmented target state vector.

Multiple models

In practice, a single dynamic model cannot always capture the real complex motion of the target because the target may move straight with a constant velocity, make turns or accelerate. In this sense, it is desirable to use the IMM algorithm [23,24,35,43] to handle multiple models of target dynamics simultaneously. In this chapter, we let M denote the number of models used to characterize the target dynamics, and $\boldsymbol{F}_m$ and $\mathbf{Q}_m$ denote the state transition matrix and process noise covariance matrix of the mth dynamic model.

2.4.2 Observation model

Multiplying (2.1a) with c and (2.1b) with c/f_o gives the expressions for the total bistatic range and range-rate measurements, respectively, for the target with respect to the transmitter and receiver i:

$$\tilde{d}_{i,k} = d_{i,k} + e_{d,i,k}, \tag{2.16a}$$

$$\tilde{\rho}_{i,k} = \rho_{i,k} + e_{\rho,i,k}, \tag{2.16b}$$

where

$$\tilde{d}_{i,k} = c\tilde{\tau}_{i,k},\ d_{i,k} = c\tau_{i,k},\ e_{d,i,k} = ce_{\tau,i,k}, \tag{2.17a}$$

$$\tilde{\rho}_{i,k} = c\tilde{\varrho}_{i,k}/f_\circ,\ \rho_{i,k} = c\varrho_{i,k}/f_\circ,\ e_{\rho,i,k} = ce_{\varrho,i,k}/f_\circ. \tag{2.17b}$$

Stacking $\tilde{d}_{i,k}$, $\tilde{\rho}_{i,k}$ and $\tilde{\theta}_{i,k}$ together forms a measurement vector at receiver i at time instant k as

$$\tilde{\boldsymbol{y}}_{i,k} = [\tilde{d}_{i,k}, \tilde{\rho}_{i,k}, \tilde{\theta}_{i,k}]^T = \boldsymbol{y}_{i,k} + \boldsymbol{e}_{i,k} = [d_{i,k}, \rho_{i,k}, \theta_{i,k}]^T + [e_{d,i,k}, e_{\rho,i,k}, e_{\theta,i,k}]^T. \tag{2.18}$$

By stacking $\tilde{\boldsymbol{y}}_{1,k}, \ldots, \tilde{\boldsymbol{y}}_{N,k}$ from all receivers, the overall measurement vector at time instant k is given by

$$\tilde{\boldsymbol{y}}_k = [\tilde{\boldsymbol{y}}_{1,k}^T, \ldots, \tilde{\boldsymbol{y}}_{N,k}^T]^T = \boldsymbol{y}_k + \boldsymbol{e}_k = [\boldsymbol{y}_{1,k}^T, \ldots, \boldsymbol{y}_{N,k}^T]^T + [\boldsymbol{e}_{1,k}^T, \ldots, \boldsymbol{e}_{N,k}^T]^T. \tag{2.19}$$

Since radar measurement errors are independent across different receivers, the covariance matrix of $\boldsymbol{e}_k$ is given by

$$\boldsymbol{R}_k = \mathbb{E}\{\boldsymbol{e}_k\boldsymbol{e}_k^T\} = \text{diag}\{\boldsymbol{R}_{1,k}, \ldots, \boldsymbol{R}_{N,k}\} \tag{2.20}$$

where the covariance matrix $\boldsymbol{R}_{i,k}$ at each receiver is computed from the corresponding CRLB matrix $\boldsymbol{C}_{i,k}$ as

$$\boldsymbol{R}_{i,k} = \boldsymbol{\Gamma}\boldsymbol{C}_{i,k}\boldsymbol{\Gamma}. \tag{2.21}$$

Here, $\boldsymbol{\Gamma}$ is defined by $\boldsymbol{\Gamma} = \text{diag}(c, c/f_\circ, 1)$ which reflects the transformations from $\{\tilde{\tau}_{i,k}, \tilde{\varrho}_{i,k}\}$ to $\{\tilde{d}_{i,k}, \tilde{\rho}_{i,k}\}$.

Note that $\boldsymbol{C}_{i,k}$ is a function of $\eta_{i,k}$ and $\boldsymbol{\psi}_k$ as discussed in Section 2.3; thus $\boldsymbol{R}_k$ is a function of $\eta_{1,k}, \ldots, \eta_{N,k}$ and $\boldsymbol{\psi}_k$. Since the main interest here is to optimize tracking performance by waveform adaptation, we only write $\boldsymbol{R}_k(\boldsymbol{\psi}_k)$ explicitly as a function of $\boldsymbol{\psi}_k$ (i.e., the transmitted waveform parameter vector).

2.4.3 Interacting multiple model – extended Kalman filter

Since the radar measurement vector $\boldsymbol{y}_k$ is a nonlinear function of the target state $\boldsymbol{x}_k$ (see (2.1), (2.2) and (2.16)–(2.19)), a nonlinear tracking algorithm must be used. Among various options available, the extended Kalman filter (EKF) is computationally most efficient compared to other more sophisticated nonlinear Kalman filtering algorithms such as the sigma-point Kalman filters including the unscented and cubature Kalman filters, and the particle filters. For the target tracking problem under consideration, the EKF can produce a good tracking performance thanks to the advantage of spatial diversity given by multistatic radar. The procedure for calculating the EKF estimates for the mth target dynamic model is given by

$$\hat{\boldsymbol{x}}_{m,k+1|k} = \boldsymbol{F}_m \hat{\boldsymbol{x}}_{m,k|k}, \tag{2.22a}$$

$$\boldsymbol{P}_{m,k+1|k} = \boldsymbol{F}_m \boldsymbol{P}_{m,k|k} \boldsymbol{F}_m^T + \boldsymbol{Q}_m, \tag{2.22b}$$

$$\boldsymbol{S}_{m,k+1}(\boldsymbol{\psi}_{k+1}) = \boldsymbol{H}_{m,k+1} \boldsymbol{P}_{m,k+1|k} \boldsymbol{H}_{m,k+1}^T + \boldsymbol{R}_{k+1}(\boldsymbol{\psi}_{k+1}), \tag{2.22c}$$

$$\boldsymbol{K}_{m,k+1}(\boldsymbol{\psi}_{k+1}) = \boldsymbol{P}_{m,k+1|k} \boldsymbol{H}_{m,k+1}^T \boldsymbol{S}_{m,k+1}^{-1}(\boldsymbol{\psi}_{k+1}), \tag{2.22d}$$

$$\tilde{\boldsymbol{z}}_{m,k+1} = \tilde{\boldsymbol{y}}_{k+1} - \boldsymbol{y}_{k+1}(\hat{\boldsymbol{x}}_{m,k+1|k}), \tag{2.22e}$$

$$\hat{\boldsymbol{x}}_{m,k+1|k+1}(\boldsymbol{\psi}_{k+1}) = \hat{\boldsymbol{x}}_{m,k+1|k} + \boldsymbol{K}_{m,k+1}(\boldsymbol{\psi}_{k+1}) \tilde{\boldsymbol{z}}_{m,k+1}, \tag{2.22f}$$

$$\boldsymbol{P}_{m,k+1|k+1}(\boldsymbol{\psi}_{k+1}) = (\boldsymbol{I} - \boldsymbol{K}_{m,k+1}(\boldsymbol{\psi}_{k+1}) \boldsymbol{H}_{m,k+1}) \boldsymbol{P}_{m,k+1|k}, \tag{2.22g}$$

where $\hat{\boldsymbol{x}}_{m,k|j}$ and $\boldsymbol{P}_{m,k|j}$ are the state estimate and error covariance, respectively, at time instant k given measurements through time instants, $0, 1, \ldots, j$, and $\boldsymbol{H}_{m,k+1} = \boldsymbol{H}_{k+1}(\hat{\boldsymbol{x}}_{m,k+1|k})$ is the Jacobian matrix $\boldsymbol{H}_{k+1}$ of $\boldsymbol{y}_{k+1}$ with respect to $\boldsymbol{x}_{k+1}$ evaluated at $\hat{\boldsymbol{x}}_{m,k+1|k}$. The expression for the Jacobian matrix $\boldsymbol{H}_{k+1}$ is provided in Section 2.8 (Appendix).

To track a maneuvering target, we employ the IMM algorithm, which can incorporate multiple models of target dynamics. The IMM algorithm is a technique of combining multiple state hypotheses by running multiple filters in parallel to obtain a better state estimate for targets with changing dynamics [23,24,35,43]. Specifically, the IMM algorithm treats the dynamic motion of the target as multiple switching models:

$$\boldsymbol{x}_{k+1} = \boldsymbol{F}(m_{k+1}) \boldsymbol{x}_k + \boldsymbol{w}_k(m_{k+1}) \tag{2.23}$$

where m_{k+1} is a finite-state Markov chain ($m_{k+1} \in \{1, \ldots, M\}$) which follows the transition probabilities p_{lm} for switching from model l to model m and the covariance matrix of the process noise $\boldsymbol{w}_k(m_{k+1})$ is governed by $\mathbb{E}\{\boldsymbol{w}_k(m_{k+1} = l)\boldsymbol{w}_k^T(m_{k+1} = l)\} = \boldsymbol{Q}(m_{k+1} = l)$ [35].

Given the model-conditioned state estimate $\hat{\boldsymbol{x}}_{m,k|k}$ and error covariance $\boldsymbol{P}_{m,k|k}$ and the model probability estimate (denoted as $\mu_{m,k|k}$) for all models $m = 1, 2 \ldots, M$ from previous time instant k, the state estimation of the IMM algorithm at time instant $k+1$ proceeds as follows.

1. Mixing of state estimates:
   - Compute the predicted model probabilities

$$\mu_{m,k+1|k} = \sum_{l=1}^{M} p_{lm} \, \mu_{l,k|k}. \tag{2.24}$$

– Compute the conditional model probabilities

$$\mu_{l|m,k|k} = \frac{1}{\mu_{m,k+1|k}} p_{lm} \, \mu_{l,k|k}. \tag{2.25}$$

– Compute the mixed estimates and covariances

$$\hat{\boldsymbol{x}}_{m^{\dagger},k|k} = \sum_{l=1}^{M} \mu_{l|m,k|k} \hat{\boldsymbol{x}}_{l,k|k}, \tag{2.26a}$$

$$\boldsymbol{P}_{m^{\dagger},k|k} = \sum_{l=1}^{M} \mu_{l|m,k|k} \left(\boldsymbol{P}_{l,k|k} + (\hat{\boldsymbol{x}}_{l,k|k} - \hat{\boldsymbol{x}}_{m^{\dagger},k|k})(\hat{\boldsymbol{x}}_{l,k|k} - \hat{\boldsymbol{x}}_{m^{\dagger},k|k})^T\right). \tag{2.26b}$$

2. Model-conditioned updates:
   - The mixed estimate $\hat{\boldsymbol{x}}_{m^{\dagger},k|k}$ and covariance $\boldsymbol{P}_{m^{\dagger},k|k}$ are used as inputs for the mth EKF to compute the state estimate $\hat{\boldsymbol{x}}_{m,k+1|k+1}$ and error covariance $\boldsymbol{P}_{m,k+1|k+1}$ at time instant $k+1$.
3. Model likelihood computations:

$$\Lambda_{m,k+1} = \left|2\pi \boldsymbol{S}_{m,k+1}\right|^{-1/2} \exp\left\{-\frac{1}{2}\tilde{\boldsymbol{z}}^T_{m,k+1}(\boldsymbol{S}_{m,k+1})^{-1}\tilde{\boldsymbol{z}}_{m,k+1}\right\} \tag{2.27}$$

where $|\cdot|$ denotes the determinant.

4. Model probability updates

$$\mu_{m,k+1|k+1} = \frac{1}{\kappa}\mu_{m,k+1|k}\Lambda_{m,k+1} \tag{2.28}$$

where κ is the normalization factor given by

$$\kappa = \sum_{l=1}^{M} \mu_{l,k+1|k}\Lambda_{l,k+1}. \tag{2.29}$$

5. Combination of state estimates:

$$\hat{\boldsymbol{x}}_{\mathrm{IMM},k+1|k+1} = \sum_{m=1}^{M} \mu_{m,k+1|k+1}\hat{\boldsymbol{x}}_{m,k+1|k+1}, \tag{2.30a}$$

$$\boldsymbol{P}_{\mathrm{IMM},k+1|k+1} = \sum_{m=1}^{M} \mu_{m,k+1|k+1}\left(\boldsymbol{P}_{m,k+1|k+1} + \Delta\hat{\boldsymbol{x}}_{m,k+1|k+1}\Delta\hat{\boldsymbol{x}}^T_{m,k+1|k+1}\right), \tag{2.30b}$$

where $\Delta\hat{\boldsymbol{x}}_{m,k+1|k+1} = \hat{\boldsymbol{x}}_{m,k+1|k+1} - \hat{\boldsymbol{x}}_{\mathrm{IMM},k+1|k+1}$.

2.5 Adaptive waveform selection

The main motivation behind adaptive waveform selection is that the tracking performance (i.e., the state estimation error covariance matrix) is dependent on the transmitted waveform parameters. To see this relationship more clearly, we recall from (2.20) that, at time instant $k+1$, the measurement covariance matrix $\boldsymbol{R}_{k+1}(\boldsymbol{\psi}_{k+1})$ is a function of the transmitted waveform parameter vector $\boldsymbol{\psi}_{k+1}$. As a result, the error covariance of filtered state estimate $\boldsymbol{P}_{m,k+1|k+1}(\boldsymbol{\psi}_{k+1})$ for the mth EKF component of the IMM-EKF algorithm is explicitly dependent on $\boldsymbol{\psi}_{k+1}$ as seen in (2.22). Consequently, the combined state error covariance matrix $\boldsymbol{P}_{\text{IMM},k+1|k+1}$ is also a function of $\boldsymbol{\psi}_{k+1}$. Based on this dependence, the tracking performance can be optimized by adaptively adjusting the parameters of the waveform transmitted at the next time instant $k+1$.

The criterion for our adaptive waveform selection scheme is to minimize the MSE of the target state estimate which is equivalent to minimizing the trace of the state estimate's error covariance matrix. Since there exist more than one error covariance matrix for the maneuvering target tracking problem under consideration, i.e., $\boldsymbol{P}_{1,k+1|k+1}, \boldsymbol{P}_{2,k+1|k+1}, \ldots, \boldsymbol{P}_{M,k+1|k+1}$ for the hypothesized models and $\boldsymbol{P}_{\text{IMM},k+1|k+1}$ for the combined state estimate, one matrix must be chosen to characterize the tracking performance. Three options are available [44]:

(i) The most-likely covariance which corresponds to the dynamic model with the highest predicted probability of being the correct model:

$$\boldsymbol{P}_{\text{Most-likely},k+1|k+1} = \boldsymbol{P}_{m^{\diamond},k+1|k+1} \tag{2.31}$$

where

$$m^{\diamond} = \arg\max_{m} \mu_{m,k+1|k}. \tag{2.32}$$

(ii) The minimax covariance which corresponds to the covariance with the largest trace value:

$$\boldsymbol{P}_{\text{Minimax},k+1|k+1} = \boldsymbol{P}_{m^{\triangleright},k+1|k+1} \tag{2.33}$$

where

$$m^{\triangleright} = \arg\max_{m} \left\{\text{trace}\left(\boldsymbol{P}_{m,k+1|k+1}\right)\right\}. \tag{2.34}$$

(iii) The predicted combined covariance matrix $\boldsymbol{P}_{\mathrm{IMM},k+1|k}$ defined by

$$\hat{\boldsymbol{x}}_{\mathrm{IMM},k+1|k} = \sum_{m=1}^{M} \mu_{m,k+1|k} \hat{\boldsymbol{x}}_{m,k+1|k}, \tag{2.35a}$$

$$\boldsymbol{P}_{\mathrm{IMM},k+1|k} = \sum_{m=1}^{M} \mu_{m,k+1|k} \left(\boldsymbol{P}_{m,k+1|k+1} + (\hat{\boldsymbol{x}}_{m,k+1|k} - \hat{\boldsymbol{x}}_{\mathrm{IMM},k+1|k})(\hat{\boldsymbol{x}}_{m,k+1|k} - \hat{\boldsymbol{x}}_{\mathrm{IMM},k+1|k})^T \right). \tag{2.35b}$$

Note that the error covariance of filtered IMM estimate, $\boldsymbol{P}_{\mathrm{IMM},k+1|k+1}$, cannot be used because the measurement vector $\tilde{\boldsymbol{y}}_{k+1}$ is not available during the waveform selection step at time instant k.

By denoting $\boldsymbol{P}^{\star}_{k+1|k+1}$ as the covariance matrix chosen from the three above covariance matrices for using in adaptive waveform selection, the waveform optimization problem is defined as

$$\boldsymbol{\psi}^{\mathrm{opt}}_{k+1} = \arg\min_{\boldsymbol{\psi} \in \boldsymbol{\Psi}} \left\{ \mathrm{trace}\left(\boldsymbol{P}^{\star}_{k+1|k+1}(\boldsymbol{\psi}) \right) \right\}, \tag{2.36}$$

where $\boldsymbol{\Psi}$ is a library of possible waveforms. The waveform library may include a number of different radar waveform classes or a single radar waveform class with various waveform parameters. Note that this waveform selection step is performed at the end of the processing at time instant k and prior to the waveform transmission at the next time instant $k+1$.

2.6 Simulation examples

Fig. 2.3 depicts the simulated tracking scenario with one transmitter located at $\boldsymbol{t} = [0, 0]^T$ m and four receivers located at $\boldsymbol{r}_1 = [20\,000, 0]^T$, $\boldsymbol{r}_2 = [10\,000, 15\,000]^T$, $\boldsymbol{r}_3 = [20\,000, -5000]^T$ and $\boldsymbol{r}_4 = [0, 10\,000]^T$ m. The target follows a nearly constant velocity trajectory with $q_x = q_y = q = 10$ m^2/s^3 from initial position $\boldsymbol{p}_0 = [27\,000, 7000]^T$ m with initial velocity $\boldsymbol{v}_0 = [-400, -200]^T$ m/s. From $t = 11$ s, the target performs a turn by uniformly changing its velocity to $[200, -200]^T$ m/s at $t = 12$ s. After the turn, the target travels with a nearly constant velocity model again with $q = 10$ m^2/s^3.

In this simulation, the transmitter emits Gaussian linear frequency-modulated (LFM) pulses. The complex envelope of LFM pulse is given by [19]

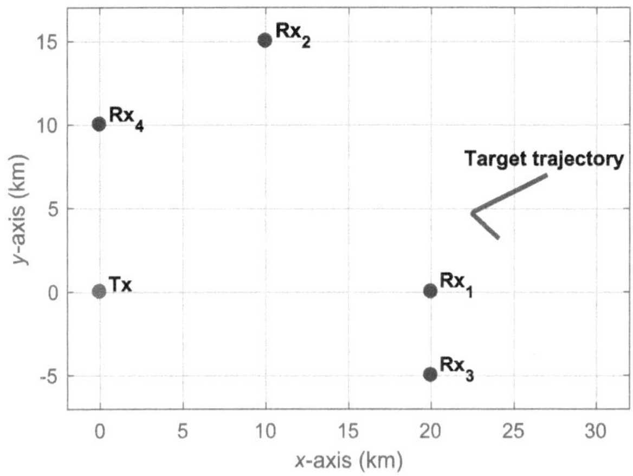

Figure 2.3 Simulated target tracking geometry.

$$\tilde{s}(t) = \left(\frac{1}{\pi\lambda^2}\right)^{1/4} \exp\left(-\left(\frac{1}{2\lambda^2} - j2\pi b\right)t^2\right) \tag{2.37}$$

where λ is the Gaussian pulse length parameter and $b = \Delta_F/(2T_s)$ is the frequency modulation rate. Here, Δ_F is the frequency sweep and T_s is the effective pulse duration which is approximated by 7.4338λ [28]. We adapt the waveform based on λ and Δ_F, i.e., the transmitted waveform parameter vector of $\boldsymbol{\psi} = [\lambda, \Delta_F]^T$, with $\lambda \in \{20, 30, 40, 50, 60\}$ μs and $\Delta_F \in \{0.1, 0.325, 0.55, 0.775, 1\}$ MHz. The waveform selection is performed via a grid search. The transmission frequency is set to $f_c = 12$ GHz, the pulse repetition interval and sampling interval to $T = 200$ m s, and the speed of signal propagation to $c = 3 \times 10^8$ m/s.

For the LFM pulses, the CRLB matrix $\boldsymbol{C}_{\{\tau,\varrho\}}$ for time delay and Doppler shift is given by [28]

$$\boldsymbol{C}_{\{\tau,\varrho\}} = \frac{1}{\eta}\begin{bmatrix} 2\lambda^2 & -4b\lambda^2 \\ -4b\lambda^2 & \dfrac{1}{2\pi^2\lambda^2} + 8b^2\lambda^2 \end{bmatrix}. \tag{2.38}$$

Therefore, the measurement error covariance matrix $\boldsymbol{R}_{i,k}$ becomes

$$\boldsymbol{R}_{i,k} = \frac{1}{\eta_{i,k}}\begin{bmatrix} 2c^2\lambda^2 & -\dfrac{4bc^2\lambda^2}{f_\circ} & 0 \\ -\dfrac{4bc^2\lambda^2}{f_\circ} & \dfrac{c^2}{f_\circ^2}\left(\dfrac{1}{2\pi^2\lambda^2} + 8b^2\lambda^2\right) & 0 \\ 0 & 0 & \sigma_\theta^2 \end{bmatrix}. \tag{2.39}$$

The SNR at receiver i at time instant k is modeled as

$$\eta_{i,k} = \frac{d_0^4}{\|\boldsymbol{p}_k - \boldsymbol{t}\|^2 \|\boldsymbol{p}_k - \boldsymbol{r}_i\|^2} \tag{2.40}$$

with $d_0 = 80\,000$ m. In the simulations, we set $\sigma_\theta = 0.04$ rad.

In this section, the IMM algorithm utilizes two nearly constant velocity models with different process noise levels to characterize hypothesized target dynamics:

Model 1: a small process noise for modeling uniform motion,

Model 2: a large process noise for modeling maneuvering motion (e.g., turning or acceleration).

Specifically, we set $q_x = q_y = q = 10$ m^2/s^3 for model 1 and $q = 100\,000$ m^2/s^3 for model 2. In a more general situation, other target dynamic models can be added to the list of hypothesized models. This extension will only require additional tracking filters to be run in parallel. The initial model probabilities are set to $\mu_{1,0|-1} = 0.9$ for model 1 and to $\mu_{2,0|-1} = 0.1$ for model 2. The transition probability matrix between models 1 and 2 is set to

$$[p_{lm}] = \begin{bmatrix} 0.9 & 0.1 \\ 0.2 & 0.8 \end{bmatrix}. \tag{2.41}$$

The track initialization is set to $\hat{\boldsymbol{x}}_{1,0|-1} = \hat{\boldsymbol{x}}_{2,0|-1} = \hat{\boldsymbol{x}}_{0|-1}$ which is drawn from a normal distribution with mean $\boldsymbol{x}_0$ and covariance $\boldsymbol{P}_{0|-1} = \text{diag}(1000^2, 1000^2, 50^2, 50^2)$.

For comparison purposes, the root mean squared error (RMSE) of the target state estimate $\hat{\boldsymbol{x}}_{\text{IMM},k|k}$ is computed from $N_{\text{MC}} = 2000$ Monte Carlo simulation runs using

$$\text{RMSE}_k = \left(\frac{1}{N_{\text{MC}}} \sum_{n=1}^{N_{\text{MC}}} \left\| \hat{\boldsymbol{x}}_{\text{IMM},k|k}^{\langle n \rangle} - \boldsymbol{x}_k^{\langle n \rangle} \right\|^2 \right)^{1/2} \tag{2.42}$$

where $\boldsymbol{x}_k^{\langle n \rangle}$ and $\hat{\boldsymbol{x}}_{\text{IMM},k|k}^{\langle n \rangle}$ denote the true and estimated target state vector at time instant k in the nth Monte Carlo run.

Adaptive waveform versus fixed waveform

Fig. 2.4 compares the RMSE performance of the adaptive waveform selection schemes presented in Section 2.5 versus two fixed waveforms with minimum and maximum time-bandwidth (TB) products. The comparison also includes the adaptive waveform selection scheme whose decision

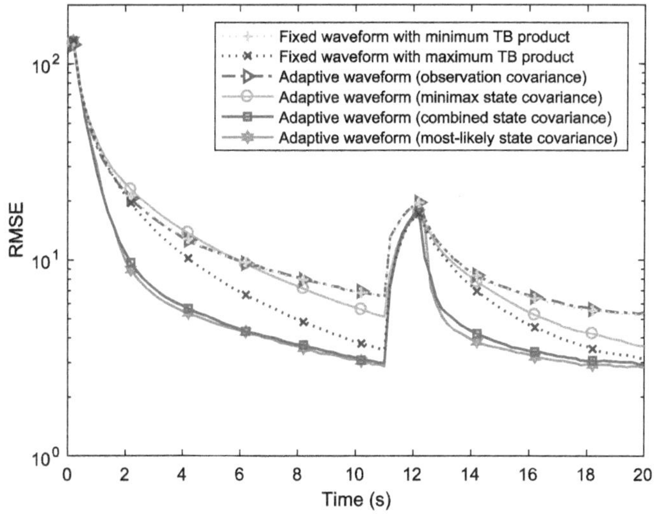

Figure 2.4 Performance comparison of various waveforms for IMM-EKF.

is made based on minimizing the trace of the observation error covariance matrix $\boldsymbol{R}_{k+1}$. Note that the performance deterioration occurring at $t = 11$–12 s is caused by the turning motion of the target during that period. We observe from Fig. 2.4 that the adaptive waveform selection schemes, which minimize the traces of the most-likely and combined covariance matrices, exhibit the best RMSE performance among the waveforms under consideration. Specifically, after the turning motion of the target, the RMSE performance of these two selection schemes recovers much faster than the other waveforms. It is interesting to observe from Fig. 2.5 that these selection schemes select alternating sequences of λ and Δ_F during the track initialization period and the track recovery period for the target turn before the selected waveforms settle down to the maximum TB waveform. In contrast, the minimum TB waveform produces the worst tracking performance. Although outperforming the minimum TB waveform, the maximum TB waveform cannot provide a tracking performance as good as the adaptive waveform selection schemes based on the most-likely and combined covariance matrices.

The adaptive waveform selection scheme that minimizes the trace of the minimax covariance matrix does not perform very well compared to the other two selection schemes mentioned above. The main reason for this is that the objective of this selection scheme is to minimize the trace of the 'largest' covariance matrix while the 'largest' covariance matrix may not

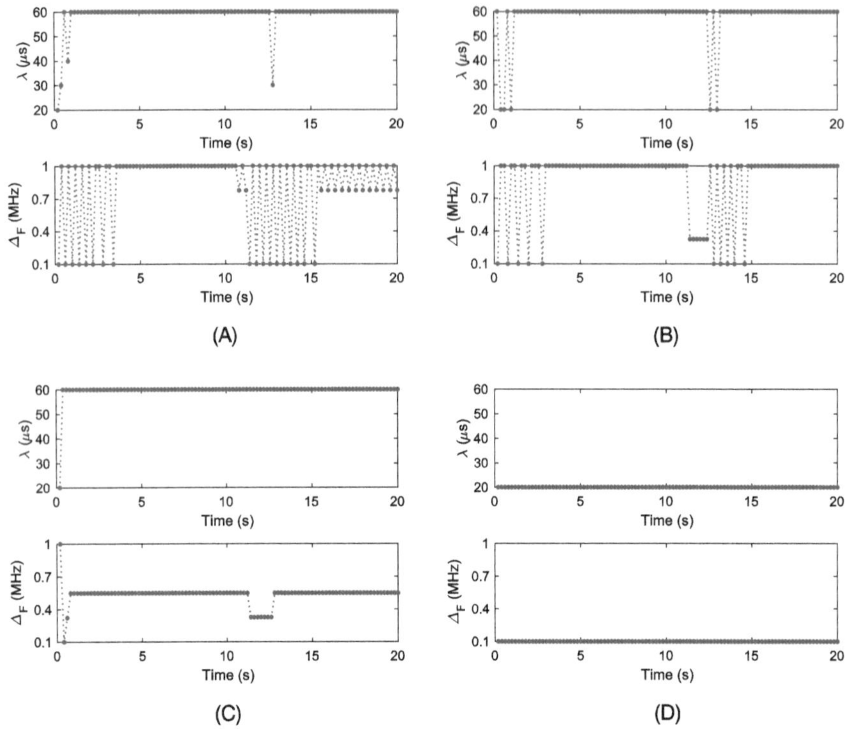

Figure 2.5 Patterns of selected waveform parameters for different adaptive waveform selection schemes. (A) Minimizing combined state covariance; (B) Minimizing most-likely state covariance; (C) Minimizing minimax state covariance; (D) Minimizing observation covariance.

be the best representative covariance matrix for the IMM algorithm. It is shown in Fig. 2.6 that the dynamic model which has the largest covariance trace often has a lower probability to be the correct model.

We also observe that the adaptive waveform selection scheme based on the observation error covariance matrix exhibits a poor RMSE performance. A similar observation was obtained in the monostatic radar setting [28]. In particular, the RMSE of this selection scheme is almost identical to that of the minimum TB waveform. This can be explained by noting that this scheme always selects the minimum TB waveform as shown in Fig. 2.5.

IMM-EKF versus EKF

We now examine the performance of the adaptive waveform selection scheme if only one EKF is utilized for target tracking. Here, the EKF makes

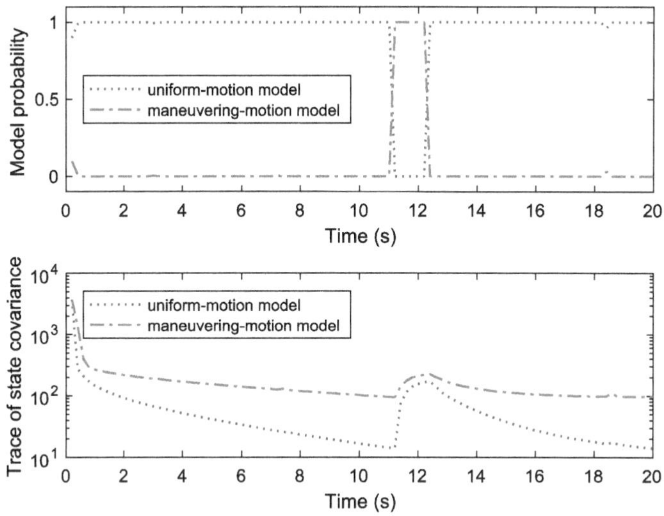

Figure 2.6 Model probabilities and traces of state covariances for the uniform-motion and maneuvering-motion models if the waveform selection is performed based on the minimax state covariance matrix.

use of a nearly constant velocity model with $q = 10$ m^2/s^3. In this case, the most-likely, minimax and combined covariance matrices become the same as only one target dynamic model is used. Fig. 2.7 plots the RMSE performance of the adaptive waveform selection method in comparison with the minimum and maximum TB waveforms. Although the adaptive waveform selection outperforms the fixed waveforms in the first ten seconds, it performs poorly after the target turn. In addition, we also observe that the alternating sequences of λ and Δ_F no longer occur during the track recovery period after the target turn as in the IMM-EKF case. We can explain this by noting the inconsistency between the RMSE of the target state estimate and the trace of the target state error covariance matrix as shown in Fig. 2.7. We can observe that the target state error covariance matrix $\boldsymbol{P}_{k|k}$ computed by the EKF does not reflect the degradation in the real RMSE performance arising from the event of target turn. Unfortunately, the waveform selection scheme decides the waveform purely based on this covariance matrix, and thus is not able to adapt the waveform to cope with the variation in target dynamics. This consequently results in a poor RMSE performance of the waveform selection scheme when a variation in target dynamics occurs. Fig. 2.8 clearly demonstrates the advantage of the use of IMM in conjunction with adaptive waveform selection.

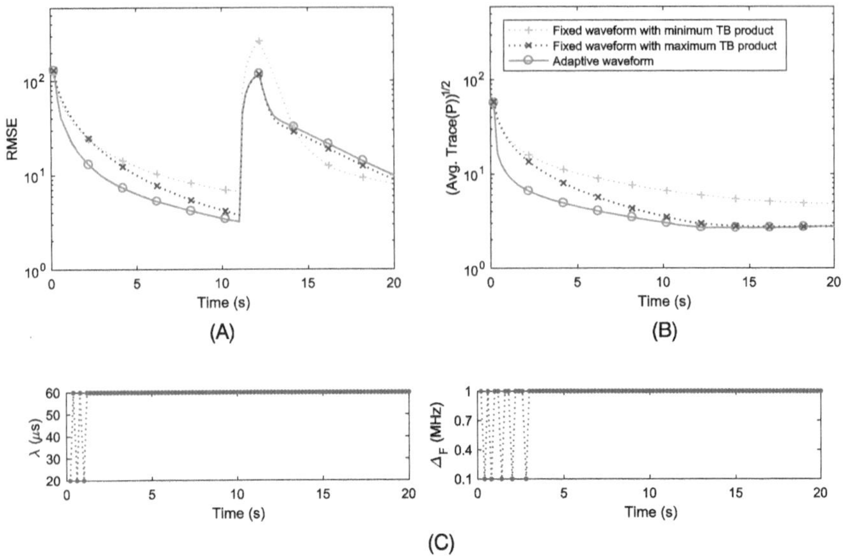

Figure 2.7 Simulation results with EKF. (A) RMSE performance; (B) Target state error covariance; (C) Patterns of selected waveform parameters.

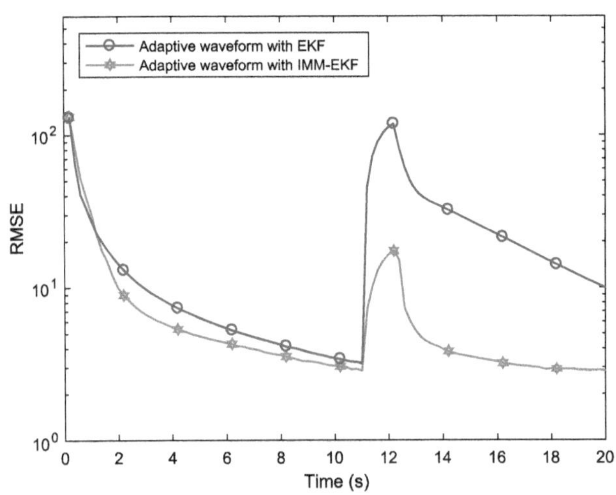

Figure 2.8 Performance comparison between IMM-EKF and EKF.

Multistatic radar versus bistatic radar

This simulation compares the RMSE performance between the multistatic radar system and four individual bistatic radars. All cases utilize the IMM-EKF algorithm jointly with the adaptive waveform selection scheme that

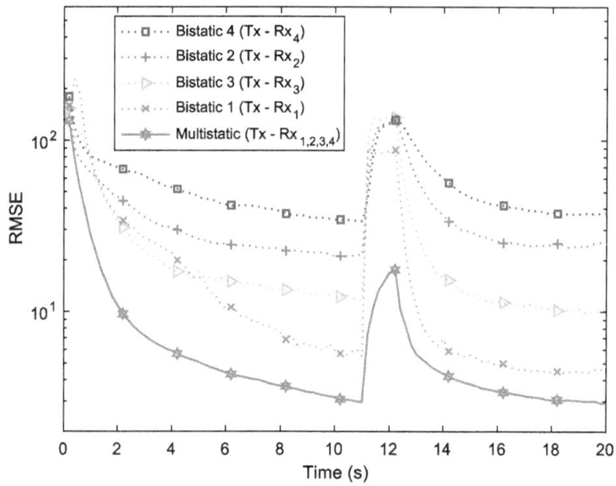

Figure 2.9 Performance comparison between multistatic and bistatic radars.

minimizes the trace of the combined covariance matrix. We observe from Fig. 2.9 that, by exploiting radar measurements from all receivers, the multistatic radar system significantly outperforms the bistatic radars, particularly during the variation in target dynamics between $t = 11$ s and 12 s.

2.7 Summary

This chapter has considered the problem of adaptive waveform selection for multistatic tracking of a maneuvering target. Three waveform selection schemes were presented based on minimizing the traces of the most-likely, minimax and combined state estimation error covariance matrices of the IMM algorithm. The performance advantages of the presented waveform selection schemes over conventional fixed waveforms were demonstrated by Monte Carlo simulation examples. The selection schemes based on the most-likely and combined covariance matrices were observed to exhibit the best RMSE performance and thus are the algorithms of choice among all the waveform schemes considered.

2.8 Appendix

The Jacobian matrix $\boldsymbol{H}_k$ of $\boldsymbol{y}_k$ with respect to $\boldsymbol{x}_k$ is

$$\boldsymbol{H}_k = [\boldsymbol{H}_{1,k}^T, \ldots, \boldsymbol{H}_{N,k}^T]^T, \tag{2.43}$$

where

$$\boldsymbol{H}_{i,k} = \begin{bmatrix} \frac{\partial d_{i,k}}{\partial p_{x,k}} & \frac{\partial d_{i,k}}{\partial p_{y,k}} & 0 & 0 \\ \frac{\partial \rho_{i,k}}{\partial p_{x,k}} & \frac{\partial \rho_{i,k}}{\partial p_{y,k}} & \frac{\partial \rho_{i,k}}{\partial v_{x,k}} & \frac{\partial \rho_{i,k}}{\partial v_{y,k}} \\ \frac{\partial \theta_{i,k}}{\partial p_{x,k}} & \frac{\partial \theta_{i,k}}{\partial p_{y,k}} & 0 & 0 \end{bmatrix} \tag{2.44}$$

for the state vector $\boldsymbol{x}_k = [\boldsymbol{p}_k^T, \boldsymbol{v}_k^T]^T$, or

$$\boldsymbol{H}_{i,k} = \begin{bmatrix} \frac{\partial d_{i,k}}{\partial p_{x,k}} & \frac{\partial d_{i,k}}{\partial p_{y,k}} & 0 & 0 & 0 & 0 \\ \frac{\partial \rho_{i,k}}{\partial p_{x,k}} & \frac{\partial \rho_{i,k}}{\partial p_{y,k}} & \frac{\partial \rho_{i,k}}{\partial v_{x,k}} & \frac{\partial \rho_{i,k}}{\partial v_{y,k}} & 0 & 0 \\ \frac{\partial \theta_{i,k}}{\partial p_{x,k}} & \frac{\partial \theta_{i,k}}{\partial p_{y,k}} & 0 & 0 & 0 & 0 \end{bmatrix} \tag{2.45}$$

for the state vector $\boldsymbol{x}_k = [\boldsymbol{p}_k^T, \boldsymbol{v}_k^T, \boldsymbol{a}_k^T]^T$. The derivative terms in (2.44) and (2.45) are given by

$$\frac{\partial d_{i,k}}{\partial p_{x,k}} = \frac{p_{x,k} - t_x}{\|\boldsymbol{p}_k - \boldsymbol{t}\|} + \frac{p_{x,k} - r_{x,i}}{\|\boldsymbol{p}_k - \boldsymbol{r}_i\|}, \tag{2.46a}$$

$$\frac{\partial d_{i,k}}{\partial p_{y,k}} = \frac{p_{y,k} - t_y}{\|\boldsymbol{p}_k - \boldsymbol{t}\|} + \frac{p_{y,k} - r_{y,i}}{\|\boldsymbol{p}_k - \boldsymbol{r}_i\|}, \tag{2.46b}$$

$$\frac{\partial \rho_{i,k}}{\partial p_{x,k}} = \frac{(p_{y,k} - t_y)(v_{x,k}(p_{y,k} - t_y) - v_{y,k}(p_{x,k} - t_x))}{\|\boldsymbol{p}_k - \boldsymbol{t}\|^3} \tag{2.46c}$$

$$+ \frac{(p_{y,k} - r_{y,i})(v_{x,k}(p_{y,k} - r_{y,i}) - v_{y,k}(p_{x,k} - r_{x,i}))}{\|\boldsymbol{p}_k - \boldsymbol{r}_i\|^3}, \tag{2.46d}$$

$$\frac{\partial \rho_{i,k}}{\partial p_{y,k}} = \frac{(p_{x,k} - t_x)(v_{y,k}(p_{x,k} - t_x) - v_{x,k}(p_{y,k} - t_y))}{\|\boldsymbol{p}_k - \boldsymbol{t}\|^3}$$

$$+ \frac{(p_{x,k} - r_{x,i})(v_{y,k}(p_{x,k} - r_{x,i}) - v_{x,k}(p_{y,k} - r_{y,i}))}{\|\boldsymbol{p}_k - \boldsymbol{r}_i\|^3}, \tag{2.46e}$$

$$\frac{\partial \rho_{i,k}}{\partial v_{x,k}} = \frac{p_{x,k} - t_x}{\|\boldsymbol{p}_k - \boldsymbol{t}\|} + \frac{p_{x,k} - r_{x,i}}{\|\boldsymbol{p}_k - \boldsymbol{r}_i\|}, \tag{2.46f}$$

$$\frac{\partial \rho_{i,k}}{\partial v_{y,k}} = \frac{p_{y,k} - t_y}{\|\boldsymbol{p}_k - \boldsymbol{t}\|} + \frac{p_{y,k} - r_{y,i}}{\|\boldsymbol{p}_k - \boldsymbol{r}_i\|}, \tag{2.46g}$$

$$\frac{\partial \theta_{i,k}}{\partial p_{x,k}} = -\frac{p_{y,k} - r_{y,i}}{\|\boldsymbol{p}_k - \boldsymbol{r}_i\|^2}, \tag{2.46h}$$

$$\frac{\partial \theta_{i,k}}{\partial p_{y,k}} = \frac{p_{x,k} - r_{x,i}}{\|\boldsymbol{p}_k - \boldsymbol{r}_i\|^2}. \tag{2.46i}$$

CHAPTER 3

Waveform selection for multistatic target tracking in clutter

Contents

3.1 Introduction and system overview

This chapter considers adaptive waveform selection for multistatic target tracking in a cluttered environment where the target detection probability is less than one and unwanted false-alarm measurements are introduced by clutter. In this chapter, we consider a multistatic radar system with one transmitter (located at position $\boldsymbol{t}$) and multiple receivers (located at positions $\boldsymbol{r}_i$, $i = 1, 2, \ldots, N$) as depicted in Fig. 3.1. Each receiver incorporates a probabilistic data association (PDA) algorithm [35] to obtain a local track. The local tracks from all receivers are communicated to the transmitter, which then performs track combining and waveform selection. The advantages of this distributed tracking scheme are twofold. First, local target tracking processes are performed in parallel by all receivers to reduce computation time. Second, the communication load between the transmitter and receivers is significantly reduced since only the local track information is transferred from the receivers to the transmitter instead of the raw measurement data, which may contain a large amount of false alarms.

For the sake of simplicity, we consider a single model for the target dynamics:

$$\boldsymbol{x}_{k+1} = \boldsymbol{F}\boldsymbol{x}_k + \boldsymbol{w}_k, \qquad \boldsymbol{w}_k \sim \mathcal{N}(\boldsymbol{0}, \mathbf{Q}) \tag{3.1}$$

even though it is straightforward to extend the problem to maneuvering target tracking using the IMM algorithm as in Chapter 2.

Assuming that the time delay and Doppler shift measurements (i.e., $\tilde{\tau}_{i,k}$ and $\tilde{\varrho}_{i,k}$) are available at each receiver i at time instant k, the total bistatic

Signal Processing for Multistatic Radar Systems
https://doi.org/10.1016/B978-0-12-815314-7.00011-1

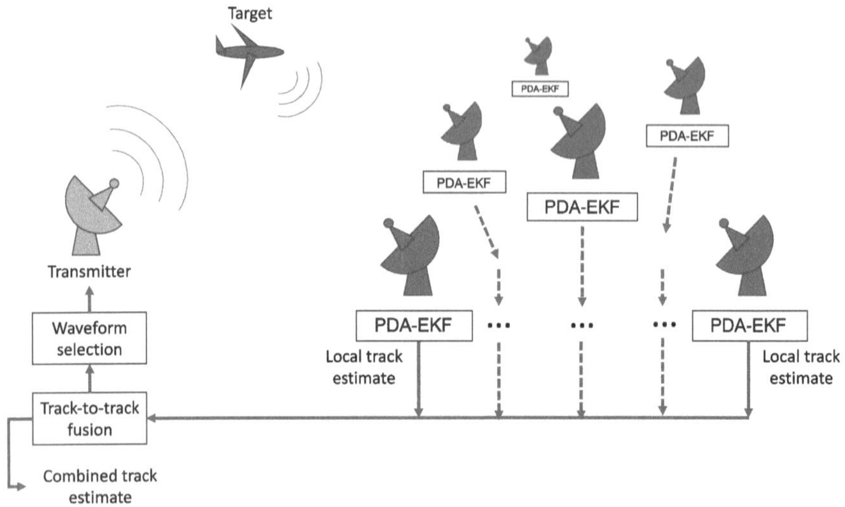

Figure 3.1 System configuration for multistatic target tracking in clutter.

range and range-rate measurements (i.e., $\tilde{d}_{i,k}$ and $\tilde{\rho}_{i,k}$) can be computed and used to form a measurement vector as

$$\tilde{\boldsymbol{y}}_{i,k} = [\tilde{d}_{i,k}, \tilde{\rho}_{i,k}]^T = \boldsymbol{y}_{i,k} + \boldsymbol{e}_{i,k} = [d_{i,k}, \rho_{i,k}]^T + [e_{d,i,k}, e_{\rho,i,k}]^T \quad (i = 1, \ldots, N) \tag{3.2}$$

where

$$d_{i,k} = \|\boldsymbol{p}_k - \boldsymbol{t}\| + \|\boldsymbol{p}_k - \boldsymbol{r}_i\|, \tag{3.3a}$$

$$\rho_{i,k} = \frac{(\boldsymbol{p}_k - \boldsymbol{t})^T \boldsymbol{v}_k}{\|\boldsymbol{p}_k - \boldsymbol{t}\|} + \frac{(\boldsymbol{p}_k - \boldsymbol{r}_i)^T \boldsymbol{v}_k}{\|\boldsymbol{p}_k - \boldsymbol{r}_i\|}. \tag{3.3b}$$

Similar to Chapter 2, we assume that the narrowband condition is satisfied and the radar measurements can approximately achieve their CRLBs under mild noise conditions. As a result, the error covariance matrix $\boldsymbol{R}_{i,k}$ is given by

$$\boldsymbol{R}_{i,k}(\boldsymbol{\psi}_k) = \boldsymbol{\Gamma}_{\{\tau,\varrho\}} \boldsymbol{C}_{\{\tau,\varrho\}}(\eta_{i,k}, \boldsymbol{\psi}_k) \boldsymbol{\Gamma}_{\{\tau,\varrho\}}, \tag{3.4}$$

where $\boldsymbol{\Gamma}_{\{\tau,\varrho\}} = \text{diag}\{c, c/f_{\circ}\}$. Although $\boldsymbol{R}_{i,k}$ is also dependent on $\eta_{i,k}$ (i.e., the SNR), it is only written explicitly as a function of the waveform parameter vector $\boldsymbol{\psi}_k$ because our focus here is on the adaptive waveform selection problem.

The probability of target detection at receiver i at time instant k is given by

$$P_{Di,k} = P_F^{1/(1+\eta_{i,k})} \tag{3.5}$$

where P_F is the probability of false alarm. The SNR, $\eta_{i,k}$, is dependent on the bistatic geometry, thus varying with time due to target motion. Note that this expression of target detection probability, originally derived for monostatic radar [30], is still valid in this case because the bistatic ambiguity function in the time delay and Doppler shift domain remains the same as the monostatic ambiguity function [39].

Along with the genuine measurements corresponding to the target (if it is detected), each receiver also picks up unwanted false measurements due to the presence of clutter. The resulting false alarms are assumed to be uniformly spatially distributed over the measurement space and independent over time. A Poisson distribution is used to model the number of false alarms, where the probability mass function of the false alarms in a volume V is given by [45]

$$\mu(m) = e^{-\varsigma V} \frac{(\varsigma V)^m}{m!} \tag{3.6}$$

where ς is the clutter density.

3.2 Tracking algorithm with probabilistic data association

Local track estimation at receivers

At each receiver i, a local track is obtained by performing the PDA-EKF algorithm using the local measurements. The PDA is a widely used technique for tracking a target in a cluttered environment. The main principle of the PDA is to utilize all the measurements that fall inside the validation region to update the track estimate based on their probability of being a target-originated measurement [35]. The EKF is used with the PDA to deal with the nonlinearity of the radar measurement $\tilde{\boldsymbol{y}}_{i,k}$ with respect to the target state $\boldsymbol{x}_k$. The PDA-EKF algorithm is summarized below.

Given the state estimate $\hat{\boldsymbol{x}}_{i,k|k}$ and its covariance $\boldsymbol{P}_{i,k|k}$ from time instant k, the predicted state estimate and covariance for time instant $k+1$ is given by

$$\hat{\boldsymbol{x}}_{i,k+1|k} = \boldsymbol{F}\hat{\boldsymbol{x}}_{i,k|k}, \tag{3.7a}$$

$$\boldsymbol{P}_{i,k+1|k} = \boldsymbol{F}\boldsymbol{P}_{i,k|k}\boldsymbol{F}^T + \boldsymbol{Q}. \tag{3.7b}$$

The validation region at time instant $k+1$ is defined as the region around the predicted measurement $\hat{\boldsymbol{y}}_{i,k+1|k}$ so that the target-originated measurement (if detected) will fall inside this region with the probability of P_G. Here, P_G is commonly known as the gating probability. Note that the predicted measurement $\hat{\boldsymbol{y}}_{i,k+1|k}$ is calculated from the predicted estimate $\hat{\boldsymbol{x}}_{i,k+1|k} = [\hat{\boldsymbol{p}}_{i,k|k+1}, \hat{\boldsymbol{v}}_{i,k|k+1}]^T$ as $\hat{\boldsymbol{y}}_{i,k+1|k} = [\hat{\tau}_{i,k+1|k}, \hat{\varrho}_{i,k+1|k}]^T$ where $\hat{\tau}_{i,k+1|k}$ and $\hat{\varrho}_{i,k+1|k}$ have the same expressions as in (3.3) except that $\boldsymbol{p}_k$ and $\boldsymbol{v}_k$ are replaced by $\hat{\boldsymbol{p}}_{i,k|k+1}$ and $\hat{\boldsymbol{v}}_{i,k|k+1}$, respectively. The condition for a point $\boldsymbol{\zeta}$ to fall inside the validation region is

$$\left(\boldsymbol{\zeta} - \hat{\boldsymbol{y}}_{i,k+1|k}\right)^T \boldsymbol{S}_{i,k+1}^{-1} \left(\boldsymbol{\zeta} - \hat{\boldsymbol{y}}_{i,k+1|k}\right) \le g^2 \tag{3.8}$$

where g is a threshold corresponding to the gating probability P_G. The threshold g is commonly known as the number of standard deviations for the validation gate and can be determined from a chi-square (χ^2) table as in [45,46]. Here, $\boldsymbol{S}_{i,k+1}$ is the residual covariance matrix at time instant $k+1$:

$$\boldsymbol{S}_{i,k+1} = \boldsymbol{H}_{i,k+1}\boldsymbol{P}_{i,k+1|k}\boldsymbol{H}_{i,k+1}^T + \boldsymbol{R}_{i,k+1} \tag{3.9}$$

where $\boldsymbol{H}_{i,k+1}$ is the Jacobian matrix of $\boldsymbol{y}_{i,k+1}$ with respect to $\boldsymbol{x}_{k+1}$ calculated using $\boldsymbol{x}_{k+1} = \hat{\boldsymbol{x}}_{i,k+1|k}$. The expression for $\boldsymbol{H}_{i,k+1}$ as a function of $\boldsymbol{x}_{k+1}$ remains the same as those given in (2.44) and (2.45) except that the last row of the matrices in (2.44) and (2.45) is removed.

We denote the number of the validated measurements within the validation region at time instant $k+1$ for the local track at receiver i as $M_{i,k+1}$. The probability for the jth validated measurement $\tilde{\boldsymbol{y}}_{i,k+1,j}$ to be a target-originated measurement is computed as

$$\beta_j = \frac{e_j}{b + \sum_{l=1}^{M_{i,k+1}} e_l} \tag{3.10}$$

for $j = 1, \ldots, M_{i,k+1}$, and the probability of having no target-originated measurement within the validation region is

$$\beta_0 = \frac{b}{b + \sum_{l=1}^{M_{i,k+1}} e_l}. \tag{3.11}$$

Here, the term e_j is given by

$$e_j = \exp\left\{-\frac{1}{2}\boldsymbol{\vartheta}_{i,k+1,j}^T \boldsymbol{S}_{i,k+1}^{-1} \boldsymbol{\vartheta}_{i,k+1,j}\right\} \tag{3.12}$$

where $\boldsymbol{\vartheta}_{i,k+1,j}$ is the residual for the jth validated measurement $\tilde{\boldsymbol{y}}_{i,k+1,j}$, i.e.,

$$\boldsymbol{\vartheta}_{i,k+1,j} = \tilde{\boldsymbol{y}}_{i,k+1,j} - \hat{\boldsymbol{y}}_{i,k+1|k}. \tag{3.13}$$

Since the number of false alarms is modeled by a Poisson distribution, b is given by

$$b = \varsigma |2\pi \boldsymbol{S}_{i,k+1}|^{1/2} \frac{(1 - P_{Di,k+1} P_G)}{P_{Di,k+1}}. \tag{3.14}$$

Note that this is the parametric version of the PDA algorithm. In the nonparametric PDA, the clutter density ς is unknown and is replaced by $M_{i,k+1}/V_{i,k+1}$, with $V_{i,k+1}$ being the hypervolume of the validation region at time instant $k+1$ at receiver i, to calculate b in (3.14).

Based on these calculated probabilities, the PDA updates the target state estimate using all the validated measurements as

$$\hat{\boldsymbol{x}}_{i,k+1|k+1} = \hat{\boldsymbol{x}}_{i,k+1|k} + \boldsymbol{K}_{i,k+1} \boldsymbol{\vartheta}_{i,k+1} \tag{3.15}$$

with the Kalman gain and combined residual given by

$$\boldsymbol{K}_{i,k+1} = \boldsymbol{P}_{i,k+1|k} \boldsymbol{H}_{i,k+1}^T \boldsymbol{S}_{i,k+1}^{-1}, \tag{3.16}$$

$$\boldsymbol{\vartheta}_{i,k+1} = \sum_{j=1}^{M_{i,k+1}} \beta_j \boldsymbol{\vartheta}_{i,k+1,j}, \tag{3.17}$$

respectively. The updated state covariance is computed as

$$\boldsymbol{P}_{i,k+1|k+1} = \beta_0 \boldsymbol{P}_{i,k+1|k} + (1 - \beta_0) \boldsymbol{P}_{i,k+1}^c + \tilde{\boldsymbol{P}}_{i,k+1} \tag{3.18}$$

where

$$\boldsymbol{P}_{i,k+1}^c = \boldsymbol{P}_{i,k+1|k} - \boldsymbol{K}_{i,k+1} \boldsymbol{S}_{i,k+1} \boldsymbol{K}_{i,k+1}^T \tag{3.19}$$

and

$$\tilde{\boldsymbol{P}}_{i,k+1} = \boldsymbol{K}_{i,k+1} \left(\sum_{j=1}^{M_{i,k+1}} \beta_j \boldsymbol{\vartheta}_{i,k+1,j} \boldsymbol{\vartheta}_{i,k+1,j}^T - \boldsymbol{\vartheta}_{i,k+1} \boldsymbol{\vartheta}_{i,k+1}^T \right) \boldsymbol{K}_{i,k+1}^T. \tag{3.20}$$

Track-to-track fusion at transmitter

Once the local track estimates $\hat{\boldsymbol{x}}_{i,k+1|k+1}$ and $\boldsymbol{P}_{i,k+1|k+1}$ from all receivers $i = 1, 2, \ldots, N$ are sent to the transmitter, a simple track-to-track fusion based on a weighted sum of the local tracks [47] is performed

$$\hat{\boldsymbol{x}}_{k+1|k+1} = \sum_{i=1}^{N} w_{i,k+1} \hat{\boldsymbol{x}}_{i,k+1|k+1}, \tag{3.21a}$$

$$\boldsymbol{P}_{k+1|k+1} = \sum_{i=1}^{N} w_{i,k+1} \boldsymbol{P}_{i,k+1|k+1}, \tag{3.21b}$$

where the weighting coefficient $w_{i,k+1}$ is defined as

$$w_{i,k+1} = \frac{\left|\boldsymbol{P}_{i,k+1|k+1}\right|^{-1}}{\sum_{l=1}^{N} \left|\boldsymbol{P}_{l,k+1|k+1}\right|^{-1}}. \tag{3.22}$$

This weighting scheme effectively gives less weight to the track with larger mean squared error. Other track-to-track fusion algorithms can also be applied here to fuse the local track estimates and their covariances [24].

The combined track estimate $\hat{\boldsymbol{x}}_{k+1|k+1}$ and its covariance $\boldsymbol{P}_{k+1|k+1}$ are then passed back to the receivers to update the local track estimates,

$$\hat{\boldsymbol{x}}_{i,k+1|k+1} = \hat{\boldsymbol{x}}_{k+1|k+1}, \tag{3.23a}$$

$$\boldsymbol{P}_{i,k+1|k+1} = \boldsymbol{P}_{k+1|k+1}, \tag{3.23b}$$

which are used for local track estimation at the receivers in the next time step.

3.3 Adaptive waveform selection

Recall from (3.4) that the measurement covariance matrix $\boldsymbol{R}_{i,k+1}(\boldsymbol{\psi}_{k+1})$ is a function of $\boldsymbol{\psi}_{k+1}$, i.e., the waveform parameter vector at time instant $k+1$. Therefore, the local track error covariance matrices $\boldsymbol{P}_{i,k+1|k+1}$ $(i = 1, \ldots, N)$ and combined track error covariance matrix $\boldsymbol{P}_{k+1|k+1}$ are also functions of $\boldsymbol{\psi}_{k+1}$. Based on this relationship, the adaptive waveform optimization problem can be formulated as

$$\boldsymbol{\psi}_{k+1}^{\text{opt}} = \arg\min_{\boldsymbol{\psi} \in \boldsymbol{\Psi}} \left\{ \text{trace}\left(\boldsymbol{P}_{k+1|k+1}(\boldsymbol{\psi})\right) \right\} \tag{3.24}$$

where $\boldsymbol{\Psi}$ is the waveform library.

Since this waveform selection step is performed prior to the actual waveform transmission at time instant $k+1$, the combined track error covariance $\boldsymbol{P}_{k+1|k+1}(\boldsymbol{\psi})$ must be calculable at the end of the processing at time instant k. However, the exact calculation of $\boldsymbol{P}_{k+1|k+1}$ requires the radar measurements at time instant $k+1$ because of the presence of false alarms. Therefore, $\boldsymbol{P}_{k+1|k+1}$ is approximated by its expected value $\bar{\boldsymbol{P}}_{k+1|k+1}$ [48], which can be computed prior to time instant $k+1$ using the modified Riccati equation.

The expected value of the local track error covariance $\boldsymbol{P}_{i,k+1|k+1}$ (denoted as $\bar{\boldsymbol{P}}_{i,k+1|k+1}$) is given by [48]

$$\boldsymbol{S}_{i,k+1}(\boldsymbol{\psi}) = \boldsymbol{H}_{i,k+1}\boldsymbol{P}_{i,k+1|k}\boldsymbol{H}_{i,k+1}^T + \boldsymbol{R}_{i,k+1}(\boldsymbol{\psi}), \tag{3.25a}$$

$$\boldsymbol{K}_{i,k+1}(\boldsymbol{\psi}) = \boldsymbol{P}_{i,k+1|k}\boldsymbol{H}_{i,k+1}^T\boldsymbol{S}_{i,k+1}^{-1}(\boldsymbol{\psi}), \tag{3.25b}$$

$$\bar{\boldsymbol{P}}_{i,k+1|k+1}(\boldsymbol{\psi}) = \boldsymbol{P}_{i,k+1|k} - q_{i,k+1}\boldsymbol{K}_{i,k+1}(\boldsymbol{\psi})\boldsymbol{S}_{i,k+1}(\boldsymbol{\psi})\boldsymbol{K}_{i,k+1}^T(\boldsymbol{\psi}). \tag{3.25c}$$

Here, $q_{i,k+1}$ is a scalar degradation factor which is dependent on the target detection probability $P_{Di,k+1}$, the clutter density ς and the validation gate volume $V_{i,k+1}$. Monte-Carlo integration can be used to evaluate $q_{i,k+1}$. However, due to the computationally demanding nature of Monte-Carlo integration, $q_{i,k+1}$ is often approximated by [49]

$$q_{i,k+1} \approx \frac{0.997\,P_{Di,k+1}}{1 + 0.37(P_{Di,k+1})^{-1.57}\varsigma\,V_{i,k+1}} \tag{3.26}$$

for the case of 2D measurement vector and 4-sigma validation gate.

The expected value of combined track error covariance is given by

$$\bar{\boldsymbol{P}}_{k+1|k+1}(\boldsymbol{\psi}) = \sum_{i=1}^{N} \bar{w}_{i,k+1}\bar{\boldsymbol{P}}_{i,k+1|k+1}(\boldsymbol{\psi}) \tag{3.27}$$

where

$$\bar{w}_{i,k+1} = \frac{\left|\bar{\boldsymbol{P}}_{i,k+1|k+1}\right|^{-1}}{\sum_{l=1}^{N}\left|\bar{\boldsymbol{P}}_{l,k+1|k+1}\right|^{-1}}. \tag{3.28}$$

By replacing $\boldsymbol{P}_{k+1|k+1}$ with $\bar{\boldsymbol{P}}_{k+1|k+1}$ in (3.24), the optimal waveform to be transmitted at next time instant $k+1$ can be selected at the end of the processing at time instant k, using

$$\boldsymbol{\psi}_{k+1}^{\text{opt}} = \arg\min_{\boldsymbol{\psi}\in\boldsymbol{\Psi}}\left\{\text{trace}\left(\bar{\boldsymbol{P}}_{k+1|k+1}(\boldsymbol{\psi})\right)\right\} \tag{3.29}$$

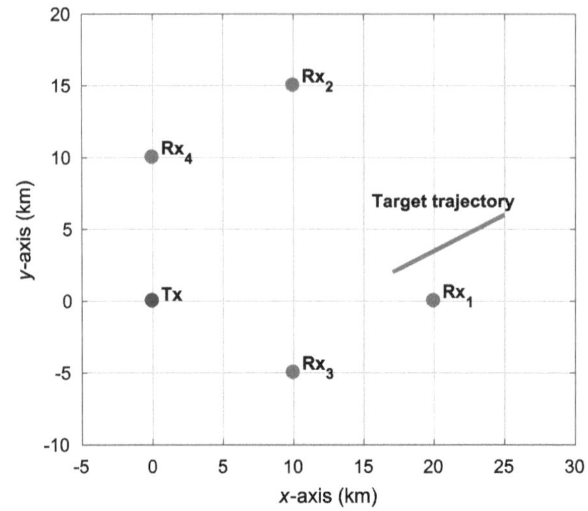

Figure 3.2 Simulated target tracking geometry.

3.4 Simulation examples

In this section, we consider a simulated tracking scenario with a transmitter and four receivers as depicted in Fig. 3.2. The transmitter is located at the origin $[0, 0]^T$ m and the receivers at $[20\,000, 0]^T$, $[10\,000, 15\,000]^T$, $[10\,000, -5000]^T$, and $[0, 10\,000]^T$ m. The target follows a nearly constant velocity dynamic model with $q_x = q_y = 10$ m^2/s^3 from initial position $[25\,000, 6000]^T$ m with initial velocity $[-400, -200]^T$ m/s. The transmission frequency is set to $f_\circ = 12$ GHz, the pulse repetition interval to $T = 200$ ms, the speed of signal propagation to $c = 3 \times 10^8$ m/s, and the probability of false alarm to $P_F = 0.01$. Here, Gaussian LFM pulses are used to construct a waveform library by varying the pulse length $\lambda \in \{10, 28, 46, 64, 82, 100\}$ μs and frequency sweep $\Delta_F \in \{0.1, 0.28, 0.46, 0.64, 0.82, 1\}$ MHz.

Substituting the CRLB matrix $\boldsymbol{C}_{\{\tau,\varrho\}}$ corresponding to the LFM pulses (2.38) into (3.4), we obtain an expression for the observation error covariance matrix $\boldsymbol{R}_{i,k}$ as

$$\boldsymbol{R}_{i,k} = \frac{1}{\eta_{i,k}} \begin{bmatrix} 2c^2\lambda^2 & -\dfrac{4bc^2\lambda^2}{f_\circ} \\ -\dfrac{4bc^2\lambda^2}{f_\circ} & \dfrac{c^2}{f_\circ^2}\left(\dfrac{1}{2\pi^2\lambda^2} + 8b^2\lambda^2\right) \end{bmatrix}. \tag{3.30}$$

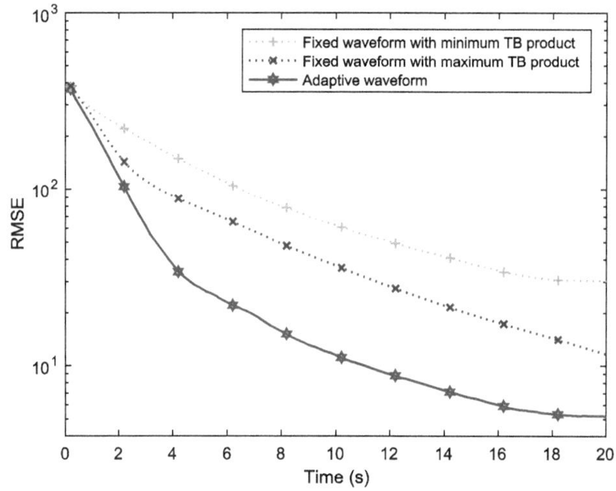

Figure 3.3 Performance comparison between adaptive and fixed waveforms for $\varsigma = 2f_\circ/c^2$.

Here, the SNR at receiver i and time instant k is modeled as $\eta_{i,k} = d_0^4/(\|\boldsymbol{p}_k - \boldsymbol{t}\|^2\|\boldsymbol{p}_k - \boldsymbol{r}_i\|^2)$ with $d_0 = 50\,000$ m. To calculate $\boldsymbol{R}_{i,k+1}$ in the waveform selection process, $\eta_{i,k+1}$ can be approximated using the predicted target–transmitter and target–receiver distances which are calculated from the predicted state estimate $\hat{\boldsymbol{x}}_{i,k+1|k}$. The track initialization $\hat{\boldsymbol{x}}_{0|-1}$ is drawn from a normal distribution with mean $\boldsymbol{x}_0$ and covariance $\boldsymbol{P}_{0|-1} = \text{diag}((100\sqrt{10})^2, (100\sqrt{10})^2, 10^2, 10^2)$. In the simulations, a 4-sigma validation gate is used, and a track is considered as lost if the correct measurement does not fall inside the validation gate in more than four consecutive time instants as in [29,30,50]. To evaluate the tracking performance, Monte Carlo simulations are carried out and the RMSE of the target state estimate $\hat{\boldsymbol{x}}_{k|k}$ is computed from 2000 converged tracks as

$$\text{RMSE}_k = \left(\frac{1}{2000} \sum_{n=1}^{2000} \left\| \hat{\boldsymbol{x}}_{k|k}^{\langle n \rangle} - \boldsymbol{x}_k^{\langle n \rangle} \right\|^2 \right)^{1/2} \tag{3.31}$$

where $\boldsymbol{x}_k^{\langle n \rangle}$ and $\hat{\boldsymbol{x}}_{k|k}^{\langle n \rangle}$ denote the true and estimated target state vector at time instant k in the nth converged track.

Figs. 3.3 and 3.4 compare the RMSE performance of the adaptive waveform selection scheme presented in Section 3.3 to the fixed minimum and maximum TB waveforms. Here, two scenarios are considered with clutter densities of $\varsigma = 2f_\circ/c^2$ and $\varsigma = 20f_\circ/c^2$ false alarms per unit gate vol-

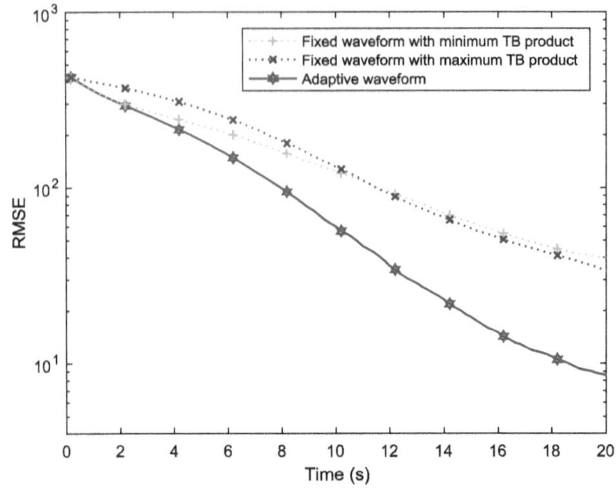

Figure 3.4 Performance comparison between adaptive and fixed waveforms for $\varsigma = 20f_\circ/c^2$.

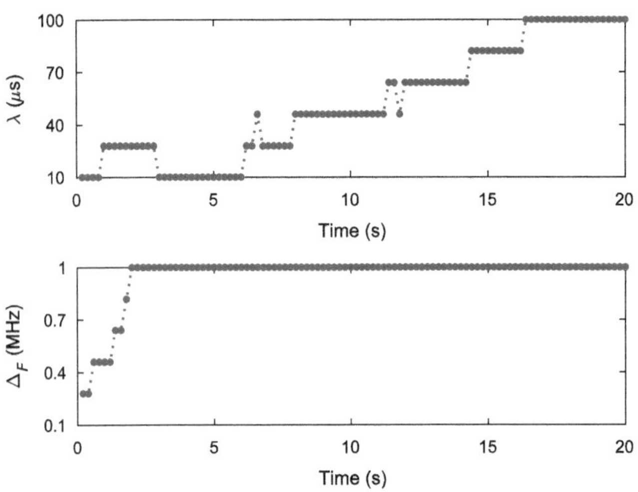

Figure 3.5 Patterns of selected waveform parameters for $\varsigma = 2f_\circ/c^2$.

ume. In both cases, the results clearly show the performance superiority of the adaptive waveform over the fixed waveforms. We also observe that the tracking performance for all waveforms degrades for a larger clutter density as expected. Patterns of the waveform parameters selected in the adaptive waveform scheme (obtained in a simulation run) is illustrated in Figs. 3.5 and 3.6.

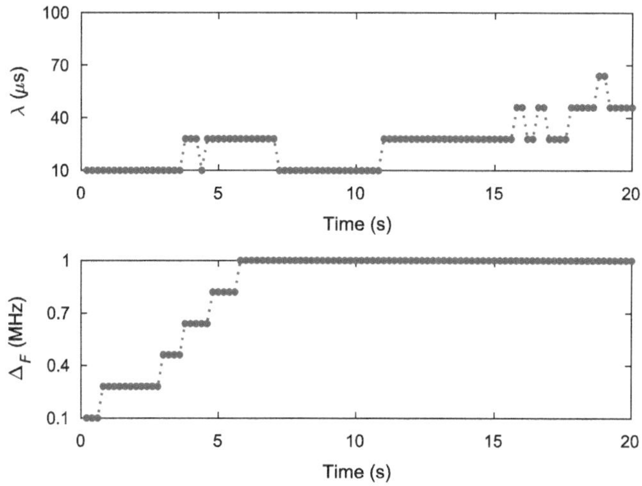

Figure 3.6 Patterns of selected waveform parameters for $\varsigma = 20f_\circ/c^2$.

It is also observed that, in order to obtain 2000 converged tracks for the first case of $\varsigma = 2f_\circ/c^2$, the adaptive waveform selection method encounters 13 track losses compared to 2 track losses for the minimum TB waveform and 40 track losses for the maximum TB waveform. In the second case with a higher clutter density $\varsigma = 20f_\circ/c^2$, the adaptive waveform, minimum TB waveform and maximum TB waveform methods experience 19, 6, and 97 track losses, respectively. This indicates that the adaptive waveform method results in a much smaller number of track losses than the worst case of using fixed maximum TB waveform in the simulated tracking scenario.

3.5 Summary

In this chapter, the optimal waveform selection problem was considered for multistatic tracking of a single target in clutter. Since the exact track error covariance matrix is not available during the waveform selection process (i.e., prior to the actual waveform transmission) due to the presence of false-alarm measurements, an expected track error covariance matrix is calculated (utilizing the degradation factor (3.26)) and used for waveform selection. It was demonstrated via simulation examples that the presented waveform selection scheme achieves a significantly reduced RMSE for the target state estimate compared to those of the fixed waveforms.

CHAPTER 4

Waveform selection for multistatic target tracking with Cartesian estimates

Contents

4.1 Introduction and system overview

Multistatic radar has a capability of using multilateration to directly estimate the target kinematic state in Cartesian coordinates based on range, Doppler and angle measurements obtained from different transmitter/receiver pairs within the system. Target tracking can thus be performed upon the available Cartesian target state estimates using a linear Kalman filter. The advantage of this approach is that it enjoys the inherent stability properties of linear estimation brought by the use of linear Kalman filter. Since the target state can be estimated in various ways via different combinations of range, Doppler and angle measurements, the available target state estimates can be combined into a single equivalent estimate before processed by the linear Kalman filter to reduce the computational load of the tracker associated with the redundant target state estimates. Another strategy is to select one target state estimate with the best accuracy for the Kalman filter input. In this chapter, the latter strategy is exploited because it does not require all receivers to communicate their measurements to the tracker, thus saving the bandwidth and power consumption in communication links within the system. Since different Cartesian state estimates have different estimation error statistics, choosing the optimal Cartesian estimate to be used is critical to the overall tracking performance of the system. Since the target–radar geometry significantly affects the accuracy of the Cartesian state estimates, it is desirable to select the Cartesian estimate in an adaptive manner depending on the current geometry of the target with respect to the radar

Signal Processing for Multistatic Radar Systems
https://doi.org/10.1016/B978-0-12-815314-7.00012-3

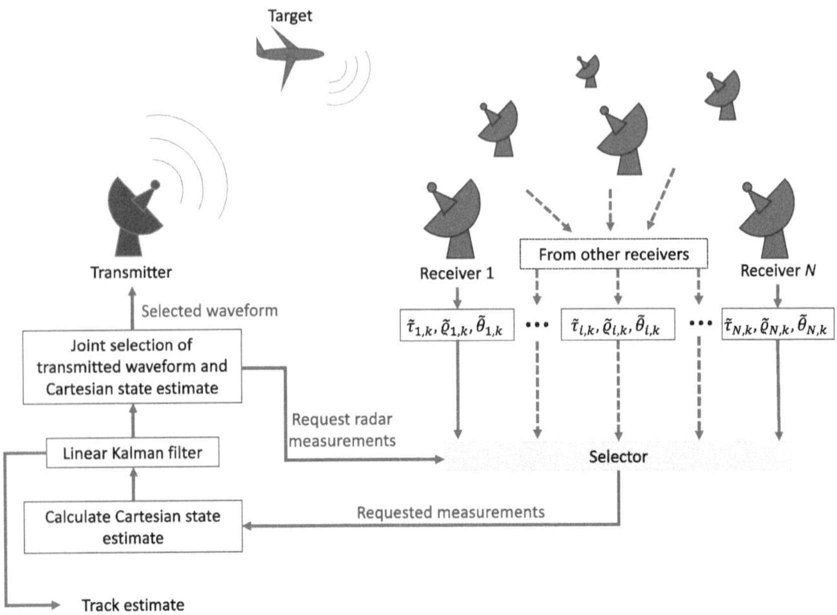

Figure 4.1 System configuration of multistatic tracking radar under consideration.

system. In this chapter, we amalgamate adaptive waveform selection and Cartesian estimate selection to form a joint selection scheme for optimizing the tracking performance.

The overall configuration of the multistatic tracking radar system under consideration is depicted in Fig. 4.1. The system consists of one transmitter located at position $\boldsymbol{t}$ and N receivers located positions $\boldsymbol{r}_i$ $(i = 1, \ldots, N)$. A fusion center, located at the transmitter site, is capable of three functions: (i) Cartesian state calculation, (ii) target tracking with a linear Kalman filter, and (iii) joint adaptive selection of radar waveform and Cartesian estimate. Each receiver is capable of obtaining time delay, Doppler shift and angle measurements. The radar waveform to be transmitted and the Cartesian estimate to be employed at each time instant, $k = 1, 2, \ldots,$ were jointly selected at the end of the previous time instant $k - 1$. After the waveform transmission at time instant k, the transmitter requests the measurements (which are required to compute the selected Cartesian estimates) from two receivers. Only these two receivers then communicate their measurements to the transmitter via their communication links with the transmitter. At the fusion center, these measurements are used to compute the Cartesian estimate which is then processed by the linear Kalman filter to track the target. Finally, radar waveform and Cartesian estimate are jointly selected again for use at the next time instant $k+1$. This tracking system offers three important advantages; viz., (i) minimal requirement for bandwidth and power consumption in transmitter-receiver communication links, (ii) performance superiority brought about by the joint adaptive selection of radar waveform and Cartesian estimate, and (iii) inherent stability of linear estimation thanks to the use of linear Kalman filter.

4.2 Target position and velocity estimation in Cartesian coordinates

Target position

Given the time delay, Doppler shift and angle measurements obtained from receiver i (i.e., $\tilde{\tau}_{i,k}$, $\tilde{\varrho}_{i,k}$ and $\tilde{\theta}_{i,k}$) and from receiver j (i.e., $\tilde{\tau}_{j,k}$, $\tilde{\varrho}_{j,k}$ and $\tilde{\theta}_{j,k}$), three options are available to uniquely determine the Cartesian position of the target: (i) using $\tilde{\tau}_{i,k}$ and $\tilde{\theta}_{i,k}$ from receiver i, (ii) using $\tilde{\tau}_{j,k}$ and $\tilde{\theta}_{j,k}$ from receiver j, or (iii) using $\tilde{\theta}_{i,k}$ from receiver i and $\tilde{\theta}_{j,k}$ from receiver j.

In the first option, the target position is determined by finding the intersection between the ellipse, which has the transmitter and receiver i at its foci with semimajor axis given by $\tilde{\tau}_{i,k}c$, and the line, that emanates from receiver i with its orientation specified by $\tilde{\theta}_{i,k}$, as depicted in Fig. 4.2(A). The intersection point is explicitly given by

$$\hat{p}_{y,k} = \begin{cases} r_{y,i} + \dfrac{\tan\tilde{\theta}_{i,k}\left((c\tilde{\tau}_{i,k})^2 - \|\boldsymbol{r}_i - \boldsymbol{t}\|^2\right)}{2\left((r_{x,i} - t_x) + \tan\tilde{\theta}_{i,k}(r_{y,i} - t_y) + c\tilde{\tau}_{i,k}\sqrt{1 + \tan^2\tilde{\theta}_{i,k}}\right)} \\ \text{if} \quad \tilde{\theta}_{i,k} \geq 0, \\ r_{y,i} + \dfrac{\tan\tilde{\theta}_{i,k}\left((c\tilde{\tau}_{i,k})^2 - \|\boldsymbol{r}_i - \boldsymbol{t}\|^2\right)}{2\left((r_{x,i} - t_x) + \tan\tilde{\theta}_{i,k}(r_{y,i} - t_y) - c\tilde{\tau}_{i,k}\sqrt{1 + \tan^2\tilde{\theta}_{i,k}}\right)} \\ \text{otherwise}, \end{cases} \tag{4.1a}$$

$$\hat{p}_{x,k} = r_{x,i} + \frac{\hat{p}_{y,k} - r_{y,i}}{\tan\tilde{\theta}_{i,k}}. \tag{4.1b}$$

Similarly, the target position estimated from $\tilde{\tau}_{j,k}$ and $\tilde{\theta}_{j,k}$ in the second option is given by

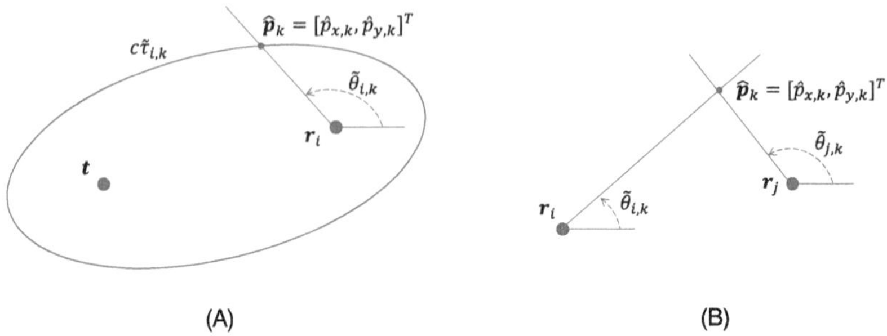

Figure 4.2 Estimation of Cartesian target position $\hat{\boldsymbol{p}}_k$. (A) Using $\tilde{\tau}_{i,k}$ and $\tilde{\theta}_{i,k}$; (B) using $\tilde{\theta}_{i,k}$ and $\tilde{\theta}_{j,k}$.

$$\hat{p}_{y,k} = \begin{cases} r_{y,j} + \dfrac{\tan\tilde{\theta}_{j,k}\left((c\tilde{\tau}_{j,k})^2 - \|\boldsymbol{r}_j - \boldsymbol{t}\|^2\right)}{2\left((r_{x,j} - t_x) + \tan\tilde{\theta}_{j,k}(r_{y,j} - t_y) + c\tilde{\tau}_{j,k}\sqrt{1 + \tan^2\tilde{\theta}_{j,k}}\right)} \\ \quad \text{if} \quad \tilde{\theta}_{j,k} \geq 0, \\ r_{y,j} + \dfrac{\tan\tilde{\theta}_{j,k}\left((c\tilde{\tau}_{j,k})^2 - \|\boldsymbol{r}_j - \boldsymbol{t}\|^2\right)}{2\left((r_{x,j} - t_x) + \tan\tilde{\theta}_{j,k}(r_{y,j} - t_y) - c\tilde{\tau}_{j,k}\sqrt{1 + \tan^2\tilde{\theta}_{j,k}}\right)} \\ \quad \text{otherwise}, \end{cases} \tag{4.2a}$$

$$\hat{p}_{x,k} = r_{x,j} + \frac{\hat{p}_{y,k} - r_{y,j}}{\tan\tilde{\theta}_{j,k}}. \tag{4.2b}$$

The third option identifies the target position by determining the intersection between the lines that pass through receiver i and j with their orientations specified by $\tilde{\theta}_{i,k}$ and $\tilde{\theta}_{j,k}$ as depicted in Fig. 4.2(B). The solution for this is given by

$$\hat{p}_{y,k} = \frac{r_{x,j} - r_{x,i} + r_{y,i}\cot\tilde{\theta}_{i,k} - r_{y,j}\cot\tilde{\theta}_{j,k}}{\cot\tilde{\theta}_{i,k} - \cot\tilde{\theta}_{j,k}}, \tag{4.3a}$$

$$\hat{p}_{x,k} = r_{x,i} + \frac{\hat{p}_{y,k} - r_{y,i}}{\tan\tilde{\theta}_{i,k}}. \tag{4.3b}$$

Target velocity

Once the target position is estimated, this information can be used in conjunction with the Doppler shift measurements $\tilde{\varrho}_{j,k}$ and $\tilde{\varrho}_{j,k}$ to calculate the Cartesian velocity of the target. Specifically, the following linear equation can be obtained by replacing $\boldsymbol{p}_k$ with $\hat{\boldsymbol{p}}_k = [\hat{p}_{x,k}, \hat{p}_{y,k}]^T$ and $\varrho_{i,k}$ with $\tilde{\varrho}_{i,k}$

in (2.2b):

$$a_{i,k}v_{x,k} + b_{i,k}v_{y,k} = c_{i,k} \tag{4.4}$$

where

$$a_{i,k} = \frac{\hat{p}_{x,k} - t_x}{\|\hat{\boldsymbol{p}}_k - \boldsymbol{t}\|} + \frac{\hat{p}_{x,k} - r_{x,i}}{\|\hat{\boldsymbol{p}}_k - \boldsymbol{r}_i\|}, \tag{4.5a}$$

$$b_{i,k} = \frac{\hat{p}_{y,k} - t_y}{\|\hat{\boldsymbol{p}}_k - \boldsymbol{t}\|} + \frac{\hat{p}_{y,k} - r_{y,i}}{\|\hat{\boldsymbol{p}}_k - \boldsymbol{r}_i\|}, \tag{4.5b}$$

$$c_{i,k} = \frac{c\tilde{\varrho}_{i,k}}{f_\circ}. \tag{4.5c}$$

Similarly for $\tilde{\varrho}_{j,k}$, we have

$$a_{j,k}v_{x,k} + b_{j,k}v_{y,k} = c_{j,k} \tag{4.6}$$

where

$$a_{j,k} = \frac{\hat{p}_{x,k} - t_x}{\|\hat{\boldsymbol{p}}_k - \boldsymbol{t}\|} + \frac{\hat{p}_{x,k} - r_{x,j}}{\|\hat{\boldsymbol{p}}_k - \boldsymbol{r}_j\|}, \tag{4.7a}$$

$$b_{j,k} = \frac{\hat{p}_{y,k} - t_y}{\|\hat{\boldsymbol{p}}_k - \boldsymbol{t}\|} + \frac{\hat{p}_{y,k} - r_{y,j}}{\|\hat{\boldsymbol{p}}_k - \boldsymbol{r}_j\|}, \tag{4.7b}$$

$$c_{j,k} = \frac{c\tilde{\varrho}_{j,k}}{f_\circ}. \tag{4.7c}$$

Using (4.4) and (4.6), the solution for $v_{x,k}$ and $v_{y,k}$ is given by

$$\hat{v}_{x,k} = \frac{c_{i,k}b_{j,k} - c_{j,k}b_{i,k}}{a_{i,k}b_{j,k} - a_{j,k}b_{i,k}}, \tag{4.8a}$$

$$\hat{v}_{y,k} = \frac{c_{i,k}a_{j,k} - c_{j,k}a_{i,k}}{b_{i,k}a_{j,k} - b_{j,k}a_{i,k}}. \tag{4.8b}$$

Target state vector

Based on the above estimation of the Cartesian position and velocity of the target, the measurements available from receiver i and j can be grouped differently to obtain three estimates of the target state vector in Cartesian coordinates:

1. $[\hat{\tau}_{i,k}, \hat{\theta}_{i,k}, \hat{\varrho}_{i,k}, \hat{\varrho}_{j,k}]^T := \boldsymbol{z}_{1,ij,k} \rightarrow$ (4.1) and (4.8) $\rightarrow [\hat{p}_{x,k}, \hat{p}_{y,k}, \hat{v}_{x,k}, \hat{v}_{y,k}]^T := \boldsymbol{y}_{1,ij,k}$.
2. $[\hat{\tau}_{j,k}, \hat{\theta}_{j,k}, \hat{\varrho}_{i,k}, \hat{\varrho}_{j,k}]^T := \boldsymbol{z}_{2,ij,k} \rightarrow$ (4.2) and (4.8) $\rightarrow [\hat{p}_{x,k}, \hat{p}_{y,k}, \hat{v}_{x,k}, \hat{v}_{y,k}]^T := \boldsymbol{y}_{2,ij,k}$.

3. $[\hat{\theta}_{i,k}, \hat{\theta}_{j,k}, \hat{\varrho}_{i,k}, \hat{\varrho}_{j,k}]^T := \boldsymbol{z}_{3,ij,k} \rightarrow$ (4.3) and (4.8) $\rightarrow [\hat{p}_{x,k}, \hat{p}_{y,k}, \hat{v}_{x,k}, \hat{v}_{y,k}]^T := \boldsymbol{y}_{3,ij,k}$.

For convenience, we denote the transformations $\boldsymbol{z}_{1,ij,k} \rightarrow \boldsymbol{y}_{1,ij,k}$, $\boldsymbol{z}_{2,ij,k} \rightarrow \boldsymbol{y}_{2,ij,k}$ and $\boldsymbol{z}_{3,ij,k} \rightarrow \boldsymbol{y}_{3,ij,k}$ as $\boldsymbol{y}_{1,ij,k} = \boldsymbol{h}_1(\boldsymbol{z}_{1,ij,k})$, $\boldsymbol{y}_{2,ij,k} = \boldsymbol{h}_2(\boldsymbol{z}_{2,ij,k})$ and $\boldsymbol{y}_{3,ij,k} = \boldsymbol{h}_3(\boldsymbol{z}_{3,ij,k})$, respectively. The Cartesian target state estimates $\boldsymbol{y}_{1,ij,k}$, $\boldsymbol{y}_{2,ij,k}$ and $\boldsymbol{y}_{3,ij,k}$ will be identical in the ideal noise-free case. However, in the presence of measurement errors, $\boldsymbol{y}_{1,ij,k}$, $\boldsymbol{y}_{2,ij,k}$ and $\boldsymbol{y}_{3,ij,k}$ are not only different in value, but also have different error statistics. The error statistics of these Cartesian target state estimates are characterized based on the method of CRLB transformation, which is discussed in the next section.

4.3 Cramér–Rao lower bounds

Following the CRLB discussion in Section 2.3 and noting that the measurement errors at receiver i are independent from those of receiver j, the CRLBs for $\boldsymbol{z}_{1,ij,k}$, $\boldsymbol{z}_{2,ij,k}$ and $\boldsymbol{z}_{3,ij,k}$ can be expressed as

$$\boldsymbol{C}_{\boldsymbol{z}_{1,ij,k}} = \begin{bmatrix} \boldsymbol{C}^{[1,1]}_{\{\tau_{i,k},\varrho_{i,k}\}} & 0 & \boldsymbol{C}^{[1,2]}_{\{\tau_{i,k},\varrho_{i,k}\}} & 0 \\ 0 & C_{\theta_{i,k}} & 0 & 0 \\ \boldsymbol{C}^{[2,1]}_{\{\tau_{i,k},\varrho_{i,k}\}} & 0 & \boldsymbol{C}^{[2,2]}_{\{\tau_{i,k},\varrho_{i,k}\}} & 0 \\ 0 & 0 & 0 & \boldsymbol{C}^{[2,2]}_{\{\tau_{j,k},\varrho_{j,k}\}} \end{bmatrix}, \tag{4.9}$$

$$\boldsymbol{C}_{\boldsymbol{z}_{2,ij,k}} = \begin{bmatrix} \boldsymbol{C}^{[1,1]}_{\{\tau_{j,k},\varrho_{j,k}\}} & 0 & 0 & \boldsymbol{C}^{[1,2]}_{\{\tau_{j,k},\varrho_{j,k}\}} \\ 0 & C_{\theta_{j,k}} & 0 & 0 \\ 0 & 0 & \boldsymbol{C}^{[2,2]}_{\{\tau_{i,k},\varrho_{i,k}\}} & 0 \\ \boldsymbol{C}^{[2,1]}_{\{\tau_{j,k},\varrho_{j,k}\}} & 0 & 0 & \boldsymbol{C}^{[2,2]}_{\{\tau_{j,k},\varrho_{j,k}\}} \end{bmatrix}, \tag{4.10}$$

$$\boldsymbol{C}_{\boldsymbol{z}_{3,ij,k}} = \mathrm{diag}\left(C_{\theta_{i,k}}, C_{\theta_{j,k}}, \boldsymbol{C}^{[2,2]}_{\{\tau_{i,k},\varrho_{i,k}\}}, \boldsymbol{C}^{[2,2]}_{\{\tau_{j,k},\varrho_{j,k}\}}\right), \tag{4.11}$$

respectively. Here, $\boldsymbol{C}^{[a,b]}_{\{\tau_{i,k},\varrho_{i,k}\}}$ denotes the (a, b)th entry of $\boldsymbol{C}_{\{\tau_{i,k},\varrho_{i,k}\}}$, and likewise for $\boldsymbol{C}^{[a,b]}_{\{\tau_{j,k},\varrho_{j,k}\}}$.

The Cartesian target state estimates $\boldsymbol{y}_{1,ij,k}$, $\boldsymbol{y}_{2,ij,k}$ and $\boldsymbol{y}_{3,ij,k}$ are essentially nonlinear transformations of $\boldsymbol{z}_{1,ij,k}$, $\boldsymbol{z}_{2,ij,k}$ and $\boldsymbol{z}_{3,ij,k}$ based on $\boldsymbol{h}_1(\boldsymbol{z}_{1,ij,k})$, $\boldsymbol{h}_2(\boldsymbol{z}_{2ij,k})$ and $\boldsymbol{h}_3(\boldsymbol{z}_{3,ij,k})$. As a result, the CRLB matrices for $\boldsymbol{y}_{1,ij,k}$, $\boldsymbol{y}_{2,ij,k}$ and $\boldsymbol{y}_{3,ij,k}$ can be computed from the CRLB matrices for $\boldsymbol{z}_{1,ij,k}$, $\boldsymbol{z}_{2,ij,k}$ and $\boldsymbol{z}_{3,ij,k}$ based on the method of CRLB transformation [22]:

$$\boldsymbol{C}_{\boldsymbol{y}_{1,ij,k}} = \frac{\partial \boldsymbol{h}_1(\boldsymbol{z}_{1,ij,k})}{\partial \boldsymbol{z}_{1,ij,k}} \boldsymbol{C}_{\boldsymbol{z}_{1,ij,k}} \left(\frac{\partial \boldsymbol{h}_1(\boldsymbol{z}_{1,ij,k})}{\partial \boldsymbol{z}_{1,ij,k}}\right)^T, \tag{4.12}$$

$$C_{y_{2,ij,k}} = \frac{\partial h_2(z_{2,ij,k})}{\partial z_{2,ij,k}} C_{z_{2,ij,k}} \left(\frac{\partial h_2(z_{2,ij,k})}{\partial z_{2,ij,k}} \right)^T, \tag{4.13}$$

$$C_{y_{3,ij,k}} = \frac{\partial h_3(z_{3,ij,k})}{\partial z_{3,ij,k}} C_{z_{3,ij,k}} \left(\frac{\partial h_3(z_{3,ij,k})}{\partial z_{3,ij,k}} \right)^T, \tag{4.14}$$

where $\partial h_1(z_{1,ij,k})/\partial z_{1,ij,k}$, $\partial h_2(z_{2,ij,k})/\partial z_{2,ij,k}$ and $\partial h_3(z_{3,ij,k})/\partial z_{3,ij,k}$ are the Jacobian matrices.

4.4 Target tracking with joint selection of radar waveform and Cartesian estimate

Target dynamics

The dynamic model for the target kinematic state vector $x_k = [p_k^T, v_k^T]^T$ is given by

$$x_{k+1} = F x_k + w_k, \quad w_k \sim \mathcal{N}(0, Q). \tag{4.15}$$

We only consider a single dynamic model here for simplicity. Extension to multiple dynamic models, e.g., for tracking a maneuvering target, is straightforward building on the use of IMM as in Chapter 2.

Observation equation

Since each pair of receivers yields three different Cartesian target state estimates as presented in Section 4.2, there exist a total of $M = 3N(N-1)/2$ Cartesian state estimates available in the system. Indexing these Cartesian estimates with $m = 1, \ldots, M$, the mth Cartesian state estimate at time instant k (denoted as $y_{m,k}$) is essentially a noisy estimate of the true target state vector x_k:

$$y_{m,k} = x_k + \epsilon_{m,k}. \tag{4.16}$$

Here, $\epsilon_{m,k}$ is the estimation error associated with $y_{m,k}$. For sufficiently small noise, the CRLB derived in Section 4.3 can be used to approximate the covariance matrix of $\epsilon_{m,k}$. Explicitly, we can write

$$R_{m,k} = \mathbb{E}\{\epsilon_{m,k}\epsilon_{m,k}^T\} \approx C_{y_{m,k}} \tag{4.17}$$

where $C_{y_{m,k}}$ is the CRLB matrix corresponding to the Cartesian state estimate $y_{m,k}$. Depending on which pair of receivers and estimation method are to be used, $C_{y_{m,k}}$ can be computed using (4.12), (4.13) or (4.14).

Target tracking with linear Kalman filter

As mentioned in Section 4.1, only one Cartesian state estimate is selected and used for the target tracking process. Let m^* denote the index of the selected Cartesian estimate. Then, the target tracking process can be performed by simply applying the linear Kalman filter:

$$\hat{\boldsymbol{x}}_{k+1|k} = \boldsymbol{F}\hat{\boldsymbol{x}}_{k|k}, \tag{4.18a}$$
$$\boldsymbol{P}_{k+1|k} = \boldsymbol{F}\boldsymbol{P}_{k|k}\boldsymbol{F}^T + \boldsymbol{Q}, \tag{4.18b}$$
$$\boldsymbol{S}_{k+1} = \boldsymbol{P}_{k+1|k} + \boldsymbol{R}_{m^*,k+1}, \tag{4.18c}$$
$$\boldsymbol{K}_{k+1} = \boldsymbol{P}_{k+1|k}\boldsymbol{S}_{k+1}^{-1}, \tag{4.18d}$$
$$\hat{\boldsymbol{x}}_{k+1|k+1} = \hat{\boldsymbol{x}}_{k+1|k} + \boldsymbol{K}_{k+1}\left(\boldsymbol{\gamma}_{m^*,k+1} - \hat{\boldsymbol{x}}_{k+1|k}\right), \tag{4.18e}$$
$$\boldsymbol{P}_{k+1|k+1} = (\boldsymbol{I} - \boldsymbol{K}_{k+1})\boldsymbol{P}_{k+1|k}. \tag{4.18f}$$

Joint optimal selection of radar waveform and Cartesian estimate

The Cartesian state estimates $\boldsymbol{\gamma}_{m,k+1}$, $m = 1, \ldots, M$, have different estimation error statistics depending on which pair of receivers is selected to supply the measurement data and which estimation method is used. Thus, it is critical to select an appropriate Cartesian estimate as the input to the linear Kalman filter to optimize the tracking performance. This is particularly important because the target–radar geometry in the multistatic setting significantly influences the accuracy of the Cartesian state estimates (see Section 4.5 for numerical illustration). Since the target–radar geometry is changing with time due to the target motion, a particular Cartesian state estimate is not always the best choice during the entire target tracking duration. Therefore, it is desirable to select the Cartesian estimate in an adaptive manner depending on the target–radar geometry. In addition to which Cartesian state estimate should be used, the tracking performance is also dependent on the radar transmitted waveform. Specifically, for the problem under consideration, the CRLB matrix $\boldsymbol{C}_{\boldsymbol{\gamma}_{m,k+1}}$ for the Cartesian estimate $\boldsymbol{\gamma}_{m,k+1}$ is a function of the transmitted waveform parameters $\boldsymbol{\psi}$ because it is computed using either $\boldsymbol{C}_{\boldsymbol{z}_{1,ij,k+1}}$ in (4.12), $\boldsymbol{C}_{\boldsymbol{z}_{2,ij,k+1}}$ in (4.13) or $\boldsymbol{C}_{\boldsymbol{z}_{3,ij,k+1}}$ in (4.14) which are functions of $\boldsymbol{\psi}$. Therefore, the covariance matrix $\boldsymbol{R}_{m,k+1}$ is also a function of $\boldsymbol{\psi}$.

To optimize the tracking performance, the selection of transmitted waveform can be performed jointly with the selection of Cartesian state estimate based on minimizing the trace of the covariance matrix of the Kalman target state estimate:

$$\{m_{k+1}^*, \boldsymbol{\psi}_{k+1}^*\} = \underset{m\in\{1,\ldots,M\},\ \boldsymbol{\psi}\in\boldsymbol{\Psi}}{\arg\min} \left\{\text{trace}\left(\boldsymbol{P}_{k+1|k+1}(m, \boldsymbol{\psi})\right)\right\} \tag{4.19}$$

where $\boldsymbol{P}_{k+1|k+1}(m, \boldsymbol{\psi})$ is defined as

$$\boldsymbol{S}_{k+1}(m, \boldsymbol{\psi}) = \boldsymbol{P}_{k+1|k} + \boldsymbol{R}_{m,k+1}(\boldsymbol{\psi}), \tag{4.20a}$$

$$\boldsymbol{K}_{k+1}(m, \boldsymbol{\psi}) = \boldsymbol{P}_{k+1|k}\boldsymbol{S}_{k+1}^{-1}(m, \boldsymbol{\psi}), \tag{4.20b}$$

$$\boldsymbol{P}_{k+1|k+1}(m, \boldsymbol{\psi}) = (\boldsymbol{I} - \boldsymbol{K}_{k+1}(m, \boldsymbol{\psi}))\boldsymbol{P}_{k+1|k}. \tag{4.20c}$$

It is important to note that $\boldsymbol{P}_{k+1|k+1}(m, \boldsymbol{\psi})$ can be computed without the radar measurements at time instant $k+1$. Thus this selection process can be performed prior to the actual radar waveform transmission at time instant $k+1$.

4.5 Simulation examples

CRLBs of Cartesian state estimates

This section illustrates numerically the dependence of the CRLBs of the Cartesian state estimates on the radar transmitted waveform and the radar–target geometry. We consider a pair of receivers, located at $\boldsymbol{r}_1 = [20\,000, 0]^T$ m and $\boldsymbol{r}_2 = [10\,000, 20\,000]^T$ m, and a transmitter located at $\boldsymbol{t} = [0, 0]^T$ m. Thee different Cartesian state estimates are $\boldsymbol{y}_1 = \boldsymbol{h}_1(\boldsymbol{z}_1)$ with $\boldsymbol{z}_1 = [\tilde{\tau}_1, \tilde{\theta}_1, \tilde{\varrho}_1, \tilde{\varrho}_2]^T$, $\boldsymbol{y}_2 = \boldsymbol{h}_2(\boldsymbol{z}_2)$ with $\boldsymbol{z}_2 = [\tilde{\tau}_2, \tilde{\theta}_2, \tilde{\varrho}_1, \tilde{\varrho}_2]^T$, and $\boldsymbol{y}_3 = \boldsymbol{h}_3(\boldsymbol{z}_3)$ with $\boldsymbol{z}_3 = [\tilde{\theta}_1, \tilde{\theta}_2, \tilde{\varrho}_1, \tilde{\varrho}_2]^T$. The CRLB matrices for these estimates are $\boldsymbol{C}_{\boldsymbol{y}_1}$, $\boldsymbol{C}_{\boldsymbol{y}_2}$ and $\boldsymbol{C}_{\boldsymbol{y}_3}$, respectively, as computed using (4.12)–(4.14). Here, the receiver indices ij and time index k have been dropped for simplicity. We assume that the transmitter emits Gaussian-LFM pulses, whose complex envelope is given in (2.37). Thus, the CRLB matrix $\boldsymbol{C}_{\{\tau_i, \varrho_i\}}$ for LFM pulses is given in (2.38). The SNR at receiver i is modeled as $\eta_i = d_0^4/(\|\boldsymbol{p} - \boldsymbol{t}\|^2 \|\boldsymbol{p} - \boldsymbol{r}_i\|^2)$ with $d_0 = 80\,000$ m. The constant σ_θ associated with the CRLB C_{θ_i} is set to 0.05 rad. The transmission frequency is set to $f_\circ = 12$ GHz, and the speed of signal propagation is $c = 3 \times 10^8$ m/s.

Figs. 4.3 and 4.4 plot the square roots of the diagonal elements of the CRLB matrices $\boldsymbol{C}_{\boldsymbol{y}_1}$, $\boldsymbol{C}_{\boldsymbol{y}_2}$ and $\boldsymbol{C}_{\boldsymbol{y}_3}$ which correspond to the lower bounds for the RMSE of the Cartesian target position and velocity estimates extracted from $\boldsymbol{y}_1$, $\boldsymbol{y}_2$ and $\boldsymbol{y}_3$. We observe from Fig. 4.3 that shorter pulse lengths lead to position estimates with higher accuracy due to the fact that the CRLB for time delay measurement is proportional to λ^2 (see (2.38)). In contrast, longer pulse lengths are observed to yield more accurate velocity estimates as consistent with the domination of $1/(2\pi^2\lambda^2)$ in the CRLB for Doppler shift measurements. The results in Fig. 4.3 confirm the independence of the CRLBs for the position components of $\boldsymbol{y}_3$ from the

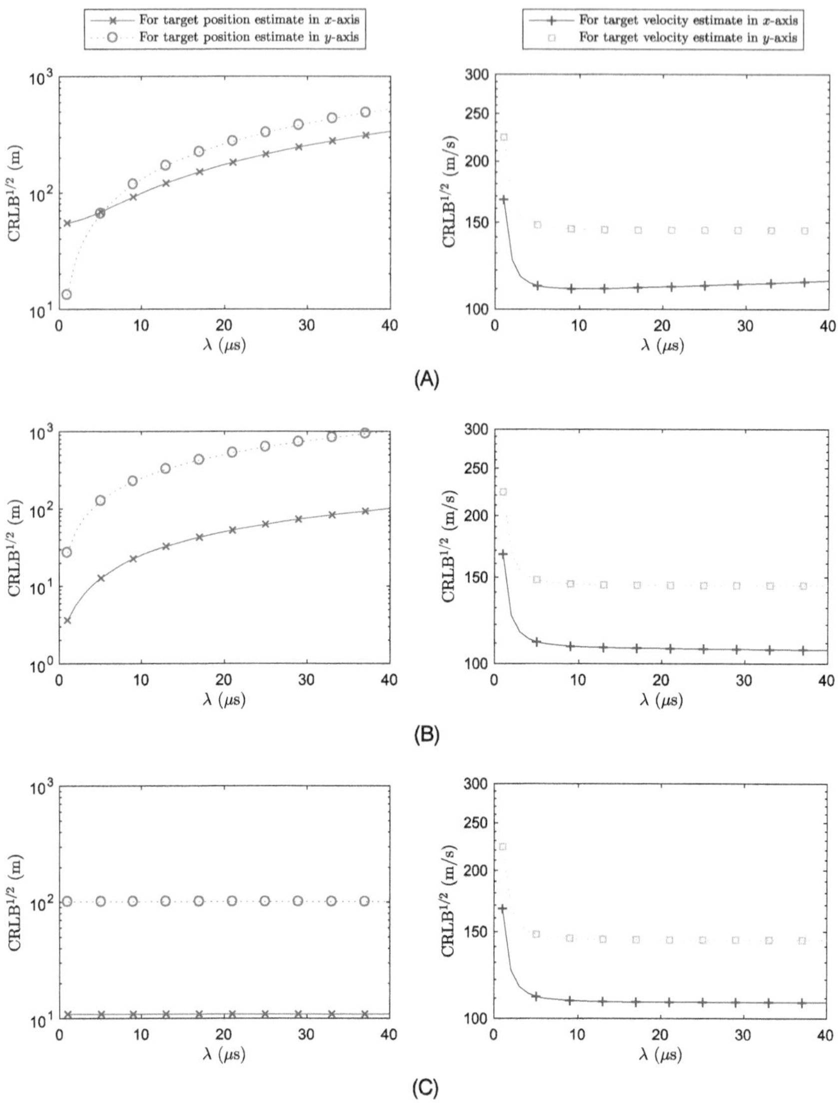

Figure 4.3 Illustration of the dependence of the CRLBs of the Cartesian state estimates on the radar transmitted waveform. The waveform frequency sweep Δ_F is set to 1 MHz. The waveform pulse length λ varies from 1 to 40 µs. (A) C_{y_1} for $y_1 = h_1(z_1)$; (B) C_{y_2} for $y_2 = h_2(z_2)$; (C) C_{y_3} for $y_3 = h_3(z_3)$.

transmitted waveform since $h_3(z_3)$ computes the Cartesian target position using $\tilde{\theta}_1$ and $\tilde{\theta}_2$ whose CRLBs do not depend on the transmitted waveform. In addition to their dependence on the transmitted waveform, the

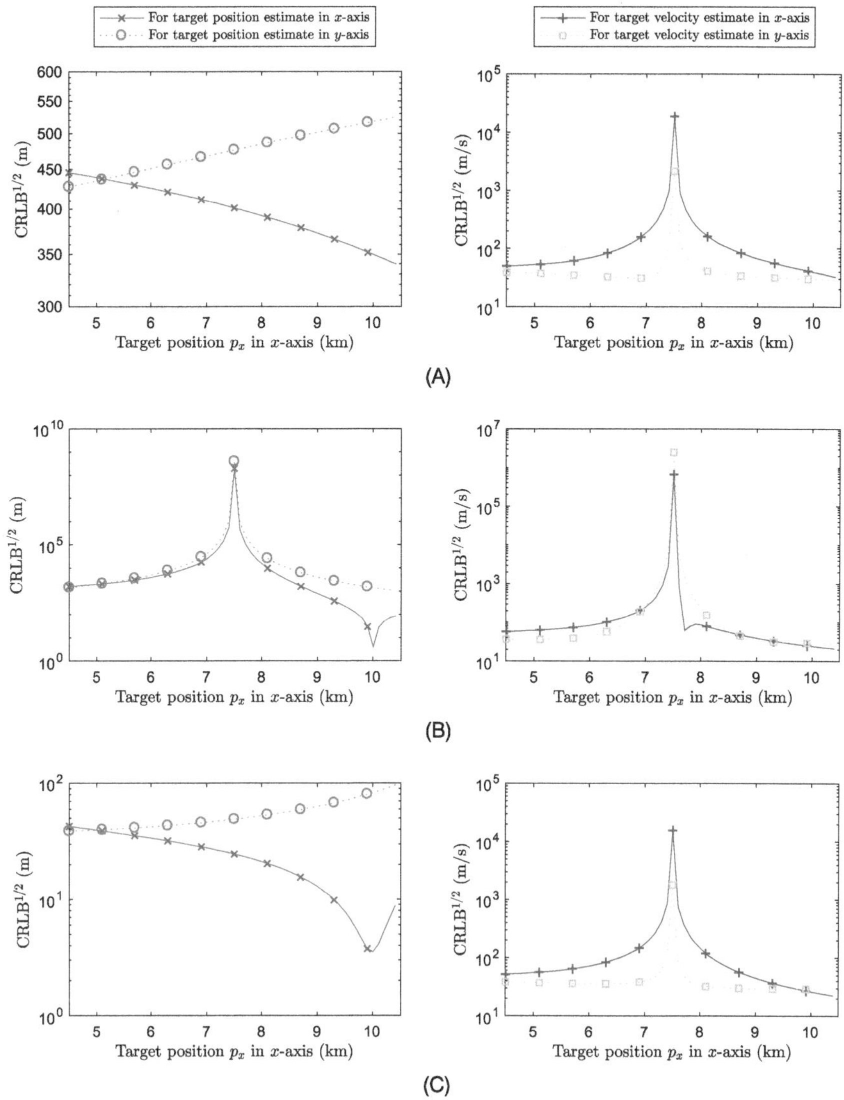

Figure 4.4 Illustration of the dependence of the CRLBs of the Cartesian state estimates on the radar–target geometry. The x-position of the target varies from 4.5 km to 10.5 km while the y-position of the target is set to 15 km. The waveform frequency sweep and pulse length are set to $\Delta_F = 0.2$ MHz and $\lambda = 40$ µs. (A) C_{y_1} for $y_1 = h_1(z_1)$; (B) C_{y_2} for $y_2 = h_2(z_2)$; (C) C_{y_3} for $y_3 = h_3(z_3)$.

CRLBs for the Cartesian estimates are observed in Fig. 4.4 to be influenced significantly by the radar–target geometry. It is interesting to note that the CRLBs for the position components of $\boldsymbol{\gamma}_2$ tend to infinity when the target approaches location $[7500, 15\,000]^T$ m. This can be explained by noting that this location lies on the baseline between the transmitter and second receiver, and thus forms a very poor bistatic geometry between the target, transmitter and second receiver. Such a behavior is also observed in the CRLBs for the velocity components of all $\boldsymbol{\gamma}_1$, $\boldsymbol{\gamma}_2$ and $\boldsymbol{\gamma}_3$ because the coefficients $a_{j,k}$ and $b_{j,k}$ in (4.6) tend to zero when the target approaches the baseline between the transmitter and second receiver, thus leading to inaccurate velocity estimation.

It is also important to observe that $\boldsymbol{C}_{\boldsymbol{\gamma}_1}$, $\boldsymbol{C}_{\boldsymbol{\gamma}_2}$ and $\boldsymbol{C}_{\boldsymbol{\gamma}_3}$ have different values, thus confirming the difference in error statistics of $\boldsymbol{\gamma}_1$, $\boldsymbol{\gamma}_2$ and $\boldsymbol{\gamma}_3$. Depending on the transmitted waveform and radar–target geometry, one of these estimates is more accurate than the others, and thus being more favorable to be used as the input to the linear Kalman filter. This highlights the importance of the selection of Cartesian estimate and transmitted waveform to the overall tracking performance of the system.

Performance advantages of joint selection of radar waveform and Cartesian estimate

We now consider a simulated tracking scenario with one transmitter located at $\boldsymbol{t} = [0, 0]^T$ m and three receivers located at $\boldsymbol{r}_1 = [20\,000, 0]^T$, $\boldsymbol{r}_2 = [10\,000, 20\,000]^T$, $\boldsymbol{r}_3 = [0, 15\,000]^T$ m as depicted in Fig. 4.5. The target follows a nearly constant velocity trajectory with $q_x = q_y = q = 10$ m^2/s^3 starting from $\boldsymbol{p}_0 = [15\,000, 10\,000]^T$ m with initial velocity $\boldsymbol{v}_0 = [-150, -100]^T$ m/s. A waveform library is constructed using the Gaussian LFM pulses by varying the pulse length $\lambda \in \{40, 50, 60, 70, 80\}$ μs and the frequency sweep $\Delta_F \in \{0.2, 0.4, 0.6, 0.8, 1\}$ MHz. The SNR at time instant k for receiver i is modeled as $\eta_{i,k} = d_0^4/(\|\boldsymbol{p}_k - \boldsymbol{t}\|^2 \|\boldsymbol{p}_k - \boldsymbol{r}_i\|^2)$ with $d_0 = 70\,000$ m. The constant σ_θ is set to 0.2 rad. In this simulation, the transmission frequency is set to $f_\circ = 12$ GHz, the pulse repetition interval to $T = 100$ ms, and the speed of signal propagation to $c = 3 \times 10^8$ m/s. Given radar measurements from three receivers, we obtain nine different Cartesian target state estimates as summarized in Table 4.1. To evaluate the tracking performance, the RMSE of the target state estimate $\hat{\boldsymbol{x}}_{k|k}$ is computed from $N_{\text{MC}} = 2000$ Monte Carlo simulation runs:

$$\text{RMSE}_k = \left(\frac{1}{N_{\text{MC}}} \sum_{n=1}^{N_{\text{MC}}} \left\| \hat{\boldsymbol{x}}_{k|k}^{(n)} - \boldsymbol{x}_k^{(n)} \right\|^2 \right)^{1/2} \tag{4.21}$$

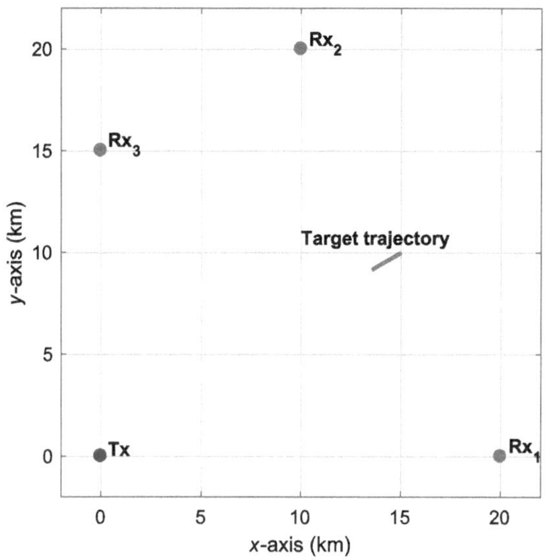

Figure 4.5 Simulated target tracking geometry.

Table 4.1 Available Cartesian target state estimates from three receivers.

Index	Receiver pair	Cartesian estimate
1	r_1, r_2	$y_{1,k} := y_{1,12,k}$
2	r_1, r_2	$y_{2,k} := y_{2,12,k}$
3	r_1, r_2	$y_{3,k} := y_{3,12,k}$
4	r_2, r_3	$y_{4,k} := y_{1,23,k}$
5	r_2, r_3	$y_{5,k} := y_{2,23,k}$
6	r_2, r_3	$y_{6,k} := y_{3,23,k}$
7	r_1, r_3	$y_{7,k} := y_{1,13,k}$
8	r_1, r_3	$y_{8,k} := y_{2,13,k}$
9	r_1, r_3	$y_{9,k} := y_{3,13,k}$

where $\boldsymbol{x}_k^{\langle n \rangle}$ and $\hat{\boldsymbol{x}}_{k|k}^{\langle n \rangle}$ denote the true and estimated target state vector at time instant k in the nth Monte Carlo run.

Fig. 4.6 shows the RMSE performance of the joint transmitted waveform and Cartesian estimate selection in comparison to that of using fixed transmitted waveform and Cartesian estimate. Here, either the maximum or minimum TB waveform is used as the fixed transmitted waveform. The fixed waveform with Cartesian estimate $y_{3,k}$ is excluded from Fig. 4.6 since the estimated track diverges in this case. This track divergence can be ex-

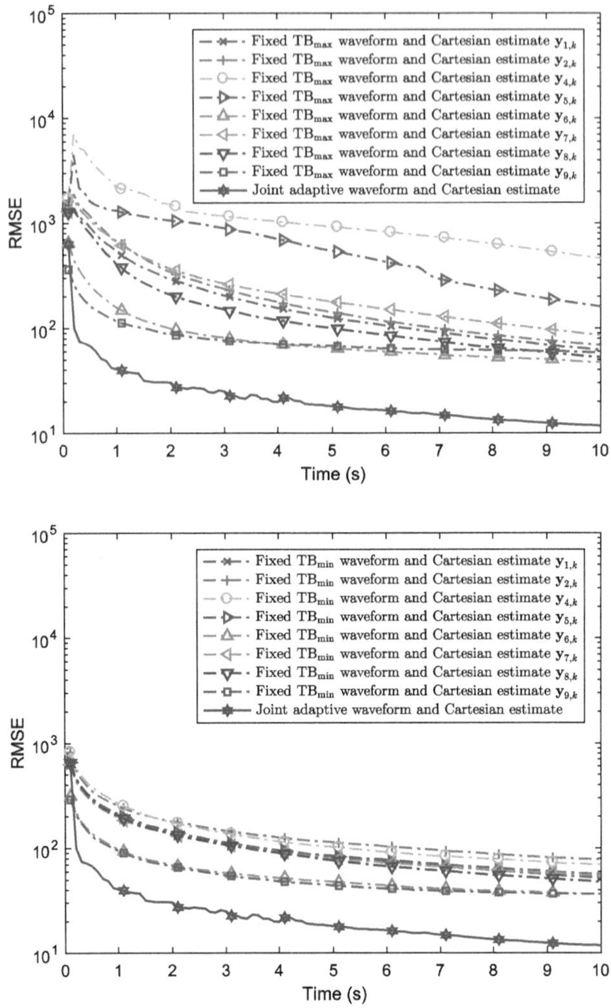

Figure 4.6 Performance comparison between the adaptive and fixed configurations of transmitted waveform and Cartesian estimate.

plained by noting the geometry of the target relatively to receivers 1 and 2. The target is located relatively close to the baseline between receivers 1 and 2, thus forming an unfavorable geometry for Cartesian estimate $\boldsymbol{y}_{3,k}$ because the position components of $\boldsymbol{y}_{3,k}$ are estimated from $\tilde{\theta}_{1,k}$ and $\tilde{\theta}_{2,k}$. The results in Fig. 4.6 evidently confirm the performance advantages of the adaptive selection of transmitted waveform and Cartesian estimate by significantly lowering the RMSE of the target state estimate. Similar to

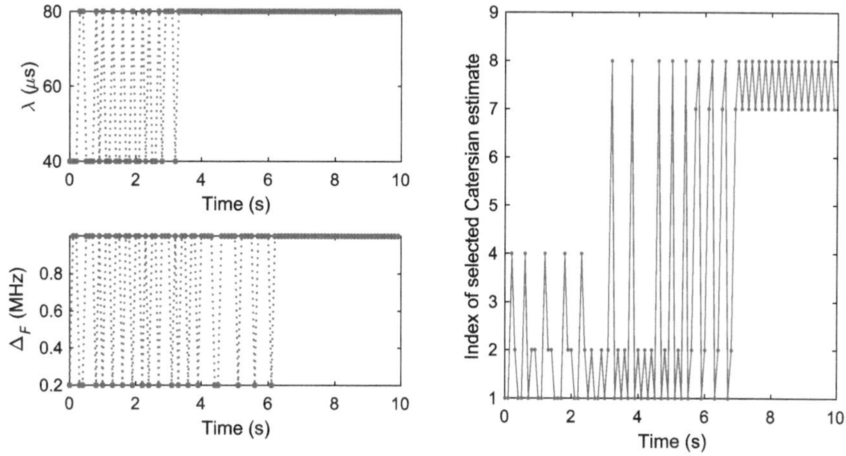

Figure 4.7 Patterns of selected waveforms and Cartesian estimates.

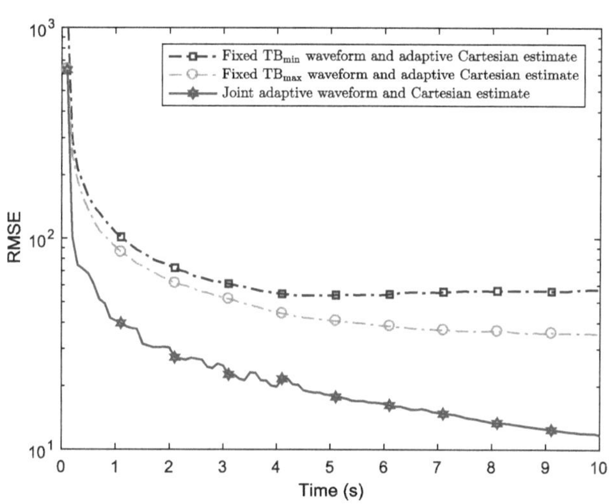

Figure 4.8 Performance comparison between the joint adaptive waveform and Cartesian estimate selection and the adaptive Cartesian estimate-only selection [51].

the observation in Section 2.6, the maximum and minimum waveform parameters are selected alternatively during the track initialization before the selected waveform settles down to the maximum TB waveform, as observed in Fig. 4.7.

Fig. 4.8 compares the RMSE performance of the joint transmitted waveform and Cartesian estimate selection against the Cartesian estimate-

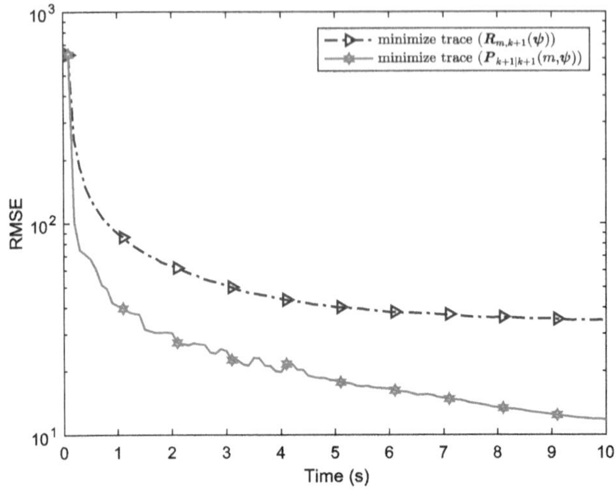

Figure 4.9 Performance comparison between selection schemes based on minimizing trace $\left(P_{k+1|k+1}(m, \psi)\right)$ and minimizing trace $\left(R_{m,k+1}(\psi)\right)$.

only selection scheme [51]. The Cartesian estimate-only selection scheme (either with the minimum or maximum TB waveform) is observed to exhibit a much larger RMSE than the joint transmitted waveform and Cartesian estimate selection, confirming the performance advantage brought by waveform diversity. In addition, the scheme in [51] selects the Cartesian target state estimate by minimizing its error covariance, which is suboptimal because minimizing the error covariance of the observation input of the Kalman filter does not necessary result in the least error covariance in the estimated track output. To further demonstrate this point, a joint selection scheme is considered for transmitted waveform and Cartesian estimate, but based on minimizing the trace of the covariance matrix of Cartesian estimation error, i.e.,

$$\{m^*_{k+1}, \psi^*_{k+1}\} = \underset{m\in\{1,\ldots,M\},\ \psi\in\Psi}{\arg\min} \left\{\text{trace}\left(R_{m,k+1}(\psi)\right)\right\} \tag{4.22}$$

It is clearly shown in Fig. 4.9 that this selection scheme results in a much larger RMSE for the Kalman target state estimates than the selection scheme based on minimizing trace$\left(P_{k+1|k+1}(m, \psi)\right)$ (i.e., the trace of the covariance matrix of the Kalman target state estimate). This is not surprising because $R_{m,k+1}(\psi)$ only contains the error statistic of the Kalman observation input at time instant $k+1$ while $P_{k+1|k+1}(m, \psi)$ incorporates

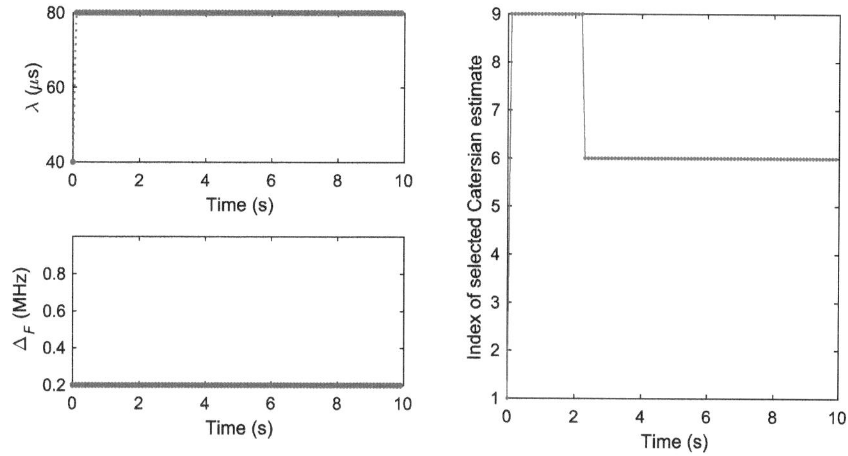

Figure 4.10 Patterns of selected waveforms and Cartesian estimates for the selection scheme based on minimizing trace$\left(\boldsymbol{R}_{m,k+1}(\boldsymbol{\psi})\right)$ in comparison with the patterns in Fig. 4.7 obtained using the selection scheme based on minimizing trace$\left(\boldsymbol{P}_{k+1|k+1}(m, \boldsymbol{\psi})\right)$.

both $\boldsymbol{R}_{m,k+1}(\boldsymbol{\psi})$ and the track information $\boldsymbol{P}_{k+1|k}$ given from the past observations. (See Fig. 4.10 for the selection pattern of the scheme (4.22).)

4.6 Summary

This chapter has developed a joint adaptive selection algorithm for both radar waveform and Cartesian estimate for multistatic tracking using Cartesian estimates. By utilizing a Cartesian target state estimate obtained via multilateration, target tracking can be performed using linear Kalman filtering with inherent stability. Since different Cartesian state estimates have different error statistics which are influenced by the radar–target geometry, choosing Cartesian estimate in an adaptive manner can provide an additional degree of freedom to enhance the performance of the overall system. In this chapter, the selection of Cartesian estimate is performed jointly with the selection of radar waveform to minimize the MSE of the Kalman target state estimate. The CRLBs for Cartesian estimates were derived and used in the development of the joint adaptive selection algorithm. The performance advantages of the algorithm were demonstrated via simulation examples.

CHAPTER 5

Waveform selection for distributed multistatic target tracking

Contents

5.1 Introduction and system overview

In previous chapters, we considered multistatic radar systems incorporating a central processor located at the transmitter site to perform target tracking as well as waveform selection. Such a centralized system configuration requires the central processor to have a high-performance computing capability to handle all signal processing for the entire system. In addition, the system becomes more vulnerable to failure and attack since all the processing power is concentrated in a single point. Depending on the system topology, the transmitter may have extensively long distances to some of the receivers. This requires high-power communication links between the transmitter and receivers, and makes the ad-hoc deployment and addition of new receivers to the system challenging.

An appealing alternative to overcome these disadvantages is to perform target tracking and waveform selection in a collaborative and fully distributed manner by all the receivers. The advantages of distributed processing over its centralized counterpart include (i) allowing parallel computing with the use of less powerful local processors, (ii) low-power communications locally between neighboring receivers, (iii) convenience of ad-hoc deployment and addition of new receivers, and (iv) robustness to link or node failure [52–54].

In this chapter, we consider a multistatic radar system with one transmitter and N receivers which form a connected network as depicted in

Signal Processing for Multistatic Radar Systems
https://doi.org/10.1016/B978-0-12-815314-7.00013-5

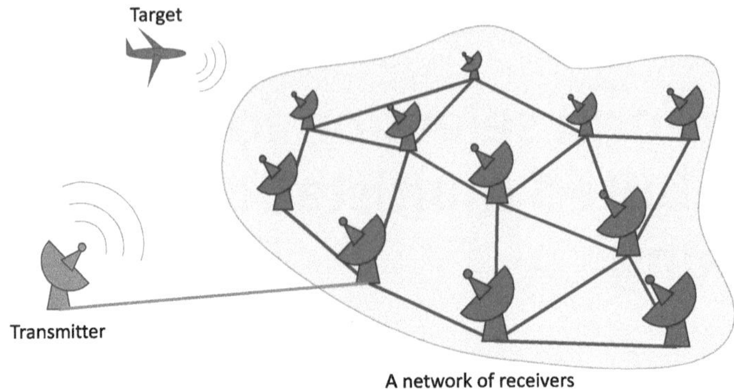

Figure 5.1 Configuration of distributed multistatic tracking radar system.

Fig. 5.1. Each receiver is linked with no hierarchy to at least one other receiver and the transmitter is connected to at least one receiver. Here, the closed neighborhood of receiver i, which includes itself and the receivers having direct links with it, is denoted by $\mathcal{V}_i$. In this configuration, the receivers share their radar measurements as well as their track estimates with their immediate neighboring receivers. Each receiver tracks the target and selects the optimal waveform by a local processor using its own information and the information transferred from its neighbors. The optimal waveform information can be transferred from any receiver to the transmitter (preferably the closest one).

We consider the following state-space model:

$$\boldsymbol{x}_{k+1} = \boldsymbol{F}\boldsymbol{x}_k + \boldsymbol{w}_k, \quad \boldsymbol{w}_k \sim \mathcal{N}(\mathbf{0}, \mathbf{Q}), \tag{5.1a}$$

$$\tilde{\boldsymbol{y}}_{i,k} = [\tilde{d}_{i,k}, \tilde{\rho}_{i,k}]^T = \boldsymbol{y}_{i,k} + \boldsymbol{e}_{i,k} = [d_{i,k}, \rho_{i,k}]^T + [e_{d,i,k}, e_{\rho,i,k}]^T \quad (i = 1, \ldots, N). \tag{5.1b}$$

Here, $\tilde{d}_{i,k}$ and $\tilde{\rho}_{i,k}$ are the total bistatic range and range-rate measurements, respectively, taken at receiver i at time instant k. For the sake of completeness, the dependence of the true bistatic range and range-rate on the true target position and velocity is given by

$$d_{i,k} = \|\boldsymbol{p}_k - \boldsymbol{t}\| + \|\boldsymbol{p}_k - \boldsymbol{r}_i\|, \tag{5.2a}$$

$$\rho_{i,k} = \frac{(\boldsymbol{p}_k - \boldsymbol{t})^T \boldsymbol{v}_k}{\|\boldsymbol{p}_k - \boldsymbol{t}\|} + \frac{(\boldsymbol{p}_k - \boldsymbol{r}_i)^T \boldsymbol{v}_k}{\|\boldsymbol{p}_k - \boldsymbol{r}_i\|}. \tag{5.2b}$$

Recall from Chapter 2 that the measurement error covariance matrix $\boldsymbol{R}_{i,k} = \mathbb{E}\left\{\boldsymbol{e}_{i,k}\boldsymbol{e}_{i,k}^T\right\}$ can be approximated using the corresponding CRLB matrix for time delay and Doppler shift $\boldsymbol{C}_{\{\tau,\varrho\}}(\eta_{i,k}, \boldsymbol{\psi}_k)$ as

$$\boldsymbol{R}_{i,k} = \boldsymbol{\Gamma}_{\{\tau,\varrho\}}\boldsymbol{C}_{\{\tau,\varrho\}}(\eta_{i,k}, \boldsymbol{\psi}_k)\boldsymbol{\Gamma}_{\{\tau,\varrho\}} \tag{5.3}$$

where $\boldsymbol{\Gamma}_{\{\tau,\varrho\}} = \mathrm{diag}\{c, c/f_\circ\}$. Since $\boldsymbol{C}_{\{\tau,\varrho\}}(\eta_{i,k}, \boldsymbol{\psi}_k)$ is related to the bistatic ambiguity function and is a function of the SNR $\eta_{i,k}$, and the transmitted waveform parameter vector $\boldsymbol{\psi}_k$, $\boldsymbol{R}_{i,k}$ is also a function of $\eta_{i,k}$ and $\boldsymbol{\psi}_k$. As we focus on the problem of optimal waveform selection, $\boldsymbol{R}_{i,k}$ will be explicitly written only as a function of $\boldsymbol{\psi}_k$, i.e., $\boldsymbol{R}_{i,k}(\boldsymbol{\psi}_k)$, in the subsequent sections.

5.2 Algorithm description

We now present an algorithm that performs target tracking (*phase A*) and selects optimal waveform (*phase B*) by all the receivers in a collaborative and distributed manner.

5.2.1 Phase A – target tracking

An extended version of the diffusion Kalman filter [53] (called DEKF) is adopted to perform distributed target tracking. The DEKF consists of two steps.

1. **Local track estimation:** After collecting the measurement data transferred from all its neighbors $l \in \mathcal{V}_i$, each receiver i computes an intermediate track estimate

$$\hat{\boldsymbol{x}}_{i,k|k-1} = \boldsymbol{F}\hat{\boldsymbol{x}}_{i,k-1|k-1}, \tag{5.4a}$$

$$\boldsymbol{P}_{i,k|k-1} = \boldsymbol{F}\boldsymbol{P}_{i,k-1|k-1}\boldsymbol{F}^T + \mathbf{Q}, \tag{5.4b}$$

$$\boldsymbol{\chi}_{i,k} = \hat{\boldsymbol{x}}_{i,k|k-1}, \tag{5.4c}$$

$$\boldsymbol{P}_{i,k} = \boldsymbol{P}_{i,k|k-1}, \tag{5.4d}$$

for every receiver $l \in \mathcal{V}_i$, repeat

$$\boldsymbol{S}_{i,k} \leftarrow \boldsymbol{H}_{l,k}\boldsymbol{P}_{i,k}\boldsymbol{H}_{l,k}^T + \boldsymbol{R}_{l,k}, \tag{5.4e}$$

$$\boldsymbol{\chi}_{i,k} \leftarrow \boldsymbol{\chi}_{i,k} + \boldsymbol{P}_{i,k}\boldsymbol{H}_{l,k}^T\boldsymbol{S}_{i,k}^{-1}(\tilde{\boldsymbol{y}}_{l,k} - \boldsymbol{y}_{l,k}(\boldsymbol{\chi}_{i,k})), \tag{5.4f}$$

$$\boldsymbol{P}_{i,k} \leftarrow \boldsymbol{P}_{i,k} - \boldsymbol{P}_{i,k}\boldsymbol{H}_{l,k}^T\boldsymbol{S}_{i,k}^{-1}\boldsymbol{H}_{l,k}\boldsymbol{P}_{i,k}, \tag{5.4g}$$

end for

$$\boldsymbol{P}_{i,k|k} = \boldsymbol{P}_{i,k}. \tag{5.4h}$$

Here, $\boldsymbol{H}_{i,k}$ is the Jacobian matrix of $\boldsymbol{y}_{i,k}$ with respect to $\boldsymbol{x}_k$ evaluated at $\hat{\boldsymbol{x}}_{i,k|k-1}$. The expression for $\boldsymbol{H}_{i,k}$ as a function of $\boldsymbol{x}_k$ remains the same as those given in (2.44) and (2.45) except that the last row of the matrices in (2.44) and (2.45) are removed.

2. **Track combination:** Once the local track estimation at all the receivers is completed, the intermediate track estimates are shared between the receivers. Each receiver i then combines all the intermediate estimates available within its neighborhood to obtain a new track estimate

$$\hat{\boldsymbol{x}}_{i,k|k} = \sum_{l \in \mathcal{V}_i} c_{i,l}\, \boldsymbol{\chi}_{l,k} \tag{5.5}$$

where $c_{i,l} = 0$ if $l \notin \mathcal{V}_i$ and $\sum_{l \in \mathcal{V}_i} c_{i,l} = 1\ \forall i$.

5.2.2 Phase B – adaptive waveform selection

Each receiver independently finds the next optimal transmitted waveform by

$$\boldsymbol{P}_{i,k+1|k} = \boldsymbol{F}\boldsymbol{P}_{i,k|k}\boldsymbol{F}^T + \boldsymbol{Q}, \tag{5.6a}$$

for every waveform $\boldsymbol{\psi} \in \boldsymbol{\Psi}$

$$\boldsymbol{P}_{i,k+1} \leftarrow \boldsymbol{P}_{i,k+1|k}, \tag{5.6b}$$

for every neighboring receiver $l \in \mathcal{V}_i$, repeat

$$\boldsymbol{S}_{i,k+1}(\boldsymbol{\psi}) \leftarrow \boldsymbol{H}_{l,k+1}\boldsymbol{P}_{i,k+1}(\boldsymbol{\psi})\boldsymbol{H}_{l,k+1}^T + \boldsymbol{R}_{l,k+1}(\boldsymbol{\psi}), \tag{5.6c}$$

$$\boldsymbol{P}_{i,k+1}(\boldsymbol{\psi}) \leftarrow \boldsymbol{P}_{i,k+1}(\boldsymbol{\psi}) - \left\{\boldsymbol{P}_{i,k+1}(\boldsymbol{\psi})\boldsymbol{H}_{l,k+1}^T\boldsymbol{S}_{i,k+1}^{-1}(\boldsymbol{\psi})\boldsymbol{H}_{l,k+1}\boldsymbol{P}_{i,k+1}(\boldsymbol{\psi})\right\}, \tag{5.6d}$$

end

$$\boldsymbol{P}_{i,k+1|k+1}(\boldsymbol{\psi}) \leftarrow \boldsymbol{P}_{i,k+1}(\boldsymbol{\psi}), \tag{5.6e}$$

end

select the waveform: $$\boldsymbol{\psi}_{i,k+1}^{\text{opt}} = \arg\min_{\boldsymbol{\psi} \in \boldsymbol{\Psi}} \left\{\text{trace}\left(\boldsymbol{P}_{i,k+1|k+1}(\boldsymbol{\psi})\right)\right\}. \tag{5.6f}$$

This scheme aims to calculate the track covariance matrix $\boldsymbol{P}_{i,k+1|k+1}$ for every single waveform candidates $\boldsymbol{\psi}$ within the waveform library $\boldsymbol{\Psi}$ and choose the waveform $\boldsymbol{\psi}_{i,k+1}^{\text{opt}}$ which yields the smallest trace value of $\boldsymbol{P}_{i,k+1|k+1}$. Since there is no fusion center in the system and the waveform selection process is performed independently at all receivers, the transmitter can make use of the waveform determined by any receiver for the next waveform transmission. Note that this distributed waveform selection scheme requires the measurement matrices $\boldsymbol{H}_{i,k+1}$ as well as the error covariance matrices $\boldsymbol{R}_{i,k+1}(\boldsymbol{\psi})$ (associated with all waveform candidates $\boldsymbol{\psi}$ within the waveform library $\boldsymbol{\Psi}$) to be shared between neighboring receivers.

5.3 Communication complexity

The total communications per receiver–receiver link required during the waveform selection phase is $2 \times 4 + ((2^2 + 2)/2) \times M = 8 + 3M$ complex scalars because $\boldsymbol{H}_{i,k+1}$ is a 2×4 matrix and $\boldsymbol{R}_{i,k+1}(\boldsymbol{\psi})$ is a 2×2 matrix. Here, M denotes the cardinality of the waveform library $\boldsymbol{\Psi}$. However, the same waveform library $\boldsymbol{\Psi}$ is used across all receivers while the CRLB matrices are functions of $\eta_{i,k+1}$ and $\boldsymbol{\psi}$. Therefore, only the SNRs, $\eta_{i,k+1}$, are required to be communicated between the neighboring receivers instead of transferring all the matrices $\boldsymbol{R}_{i,k+1}(\boldsymbol{\psi})$ corresponding to all waveform candidates $\boldsymbol{\psi}$. In addition, since $\boldsymbol{H}_{i,k+1}$ is also used in the target tracking phase of time instant $k+1$, sharing this information in the waveform selection phase at time instant k obviates sharing it in the target tracking phase at time instant $k+1$. As a result, the presented algorithm requires only one extra complex scalar, which is the SNR information, $\eta_{i,k+1}$, to be communicated via each link in comparison to the diffusion-based distributed target tracking algorithm without waveform selection.

5.4 Simulation examples

Fig. 5.2 depicts the simulated tracking scenario with a distributed multistatic radar system consisting of a transmitter and seven receivers, showing the system geometry as well as the network topology between the receivers. The target follows a nearly constant velocity dynamics with $\boldsymbol{x}_0 = [25\,000 \text{ m}, 15\,000 \text{ m}, -400 \text{ m/s}, -200 \text{ m/s}]^T$ and $q_x = q_y = q = 10 \text{ m}^2/\text{s}^3$. In this simulation, the transmitter uses Gaussian LFM pulses and a waveform library is formed by varying the pulse length λ and frequency sweep Δ_F of LFM pulse. Specifically, we have $\lambda \in \{10, 28, 46, 64, 82, 100\}$ μs and $\Delta_F \in \{0.1, 0.28, 0.46, 0.64, 0.82, 1\}$ MHz. The transmission frequency is set to $f_o = 12$ GHz, the pulse repetition interval to $T = 200$ ms, and the speed of signal propagation is $c = 3 \times 10^8$ m/s.

Substituting the CRLB matrix $\boldsymbol{C}_{\{\tau,\varrho\}}$ corresponding to the LFM pulses in (2.38) into (5.3), we obtain an expression for the observation error co-

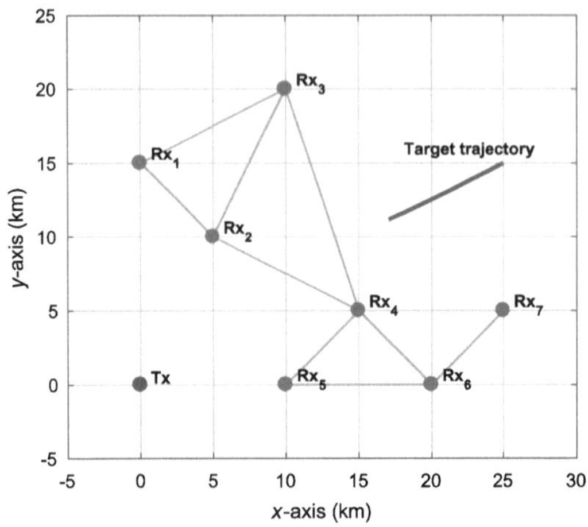

Figure 5.2 Simulated tracking geometry and network topology.

variance matrix $\boldsymbol{R}_{i,k}$ as

$$\boldsymbol{R}_{i,k} = \frac{1}{\eta_{i,k}} \begin{bmatrix} 2c^2\lambda^2 & -\dfrac{4bc^2\lambda^2}{f_\circ} \\ -\dfrac{4bc^2\lambda^2}{f_\circ} & \dfrac{c^2}{f_\circ^2}\left(\dfrac{1}{2\pi^2\lambda^2} + 8b^2\lambda^2\right) \end{bmatrix}. \tag{5.7}$$

Here, the SNR at receiver i and time instant k is modeled as $\eta_{i,k} = d_0^4/(\|\boldsymbol{p}_k - \boldsymbol{t}\|^2\|\boldsymbol{p}_k - \boldsymbol{r}_i\|^2)$ with $d_0 = 30\,000$ m. To calculate $\boldsymbol{R}_{i,k+1}(\boldsymbol{\psi})$ in the waveform selection phase, $\eta_{i,k+1}$ can be approximated using the predicted target-transmitter and target-receiver distances which are calculated from the predicted state estimate $\hat{\boldsymbol{x}}_{i,k+1|k}$. The track initialization $\hat{\boldsymbol{x}}_{0|-1}$ is drawn from a normal distribution with mean $\boldsymbol{x}_0$ and covariance $\boldsymbol{P}_{0|-1} = \text{diag}\{1000^2, 1000^2, 50^2, 50^2\}$. The combination coefficients $\{c_{l,i}\}$ for the DEKF are weighted according to how many neighbors each receiver is linked to [53].

For comparison purposes, the averaged RMSE is obtained via Monte Carlo simulations, defined as

$$\text{RMSE}_k = \left(\frac{1}{7N_{\text{MC}}} \sum_{n=1}^{N_{\text{MC}}} \sum_{i=1}^{7} \left\| \hat{\boldsymbol{x}}_{i,k|k}^{\langle n \rangle} - \boldsymbol{x}_k^{\langle n \rangle} \right\|^2 \right)^{1/2} \tag{5.8}$$

where $\boldsymbol{x}_k^{\langle n \rangle}$ denotes the true target state vector at time instant k in the nth Monte Carlo run and $\hat{\boldsymbol{x}}_{i,k|k}^{\langle n \rangle}$ denote the estimate of $\boldsymbol{x}_k^{\langle n \rangle}$ calculated at re-

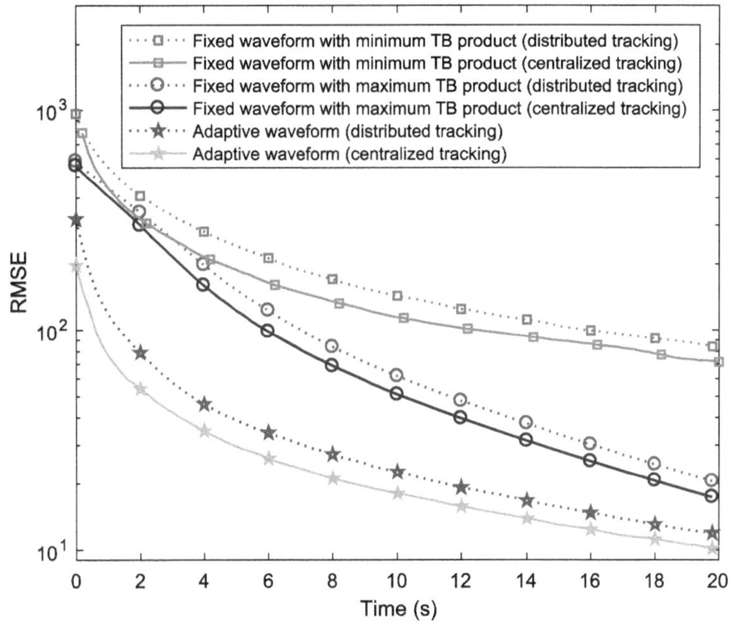

Figure 5.3 Performance comparison of various waveform configurations.

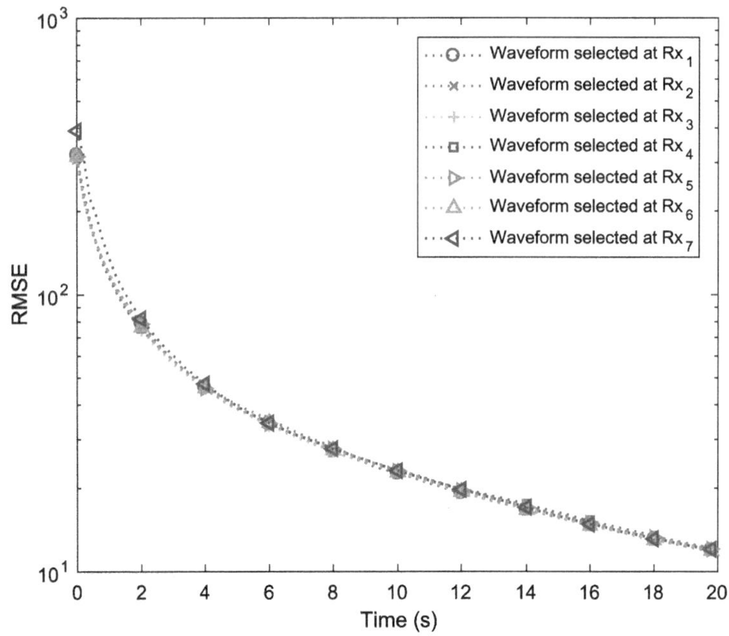

Figure 5.4 Performance comparison of waveforms selected different receivers.

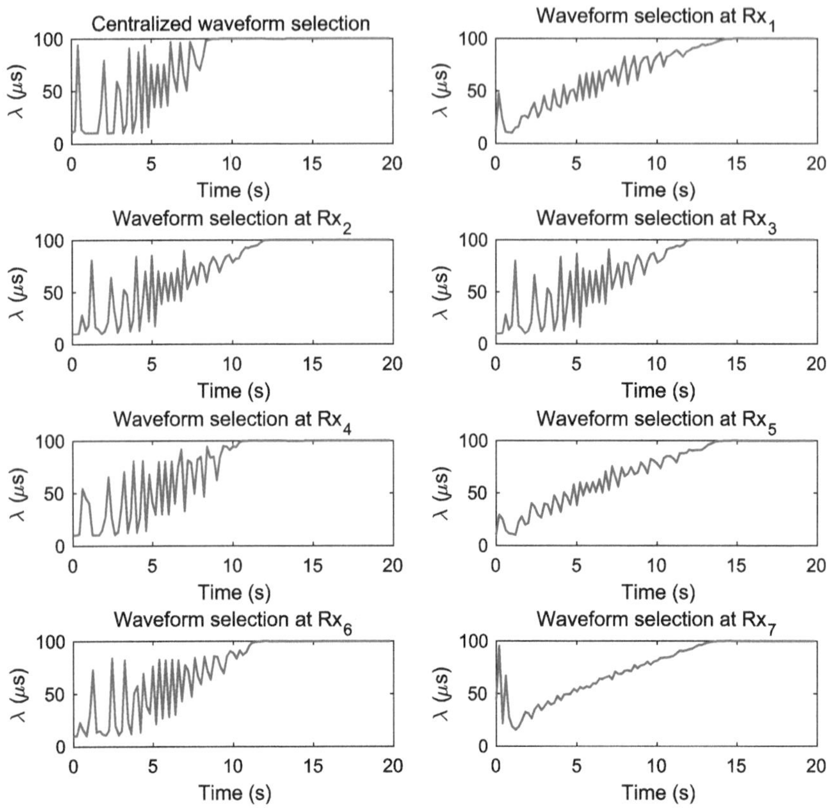

Figure 5.5 Patterns of selected waveform pulse lengths.

ceiver i. Here, $N_{MC} = 2000$ is the total number of Monte Carlo runs carried out.

Fig. 5.3 compares the RMSE performance of the distributed target tracking and waveform selection algorithm presented in Section 5.2 (transmitting the waveform selected by receiver 1 at the transmitter) with that of the DEKF algorithm using a fixed minimum TB waveform or maximum TB waveform. Here, the comparison also includes the centralized counterparts of these algorithms as performance benchmarks. The centralized algorithms assume that all receivers can communicate their measurements to the transmitter and both the target tracking and waveform selection processes are performed in a centralized manner at the transmitter. We observe that the distributed algorithms exhibit a RMSE performance quite close to their corresponding centralized counterparts. At the expense of a minor

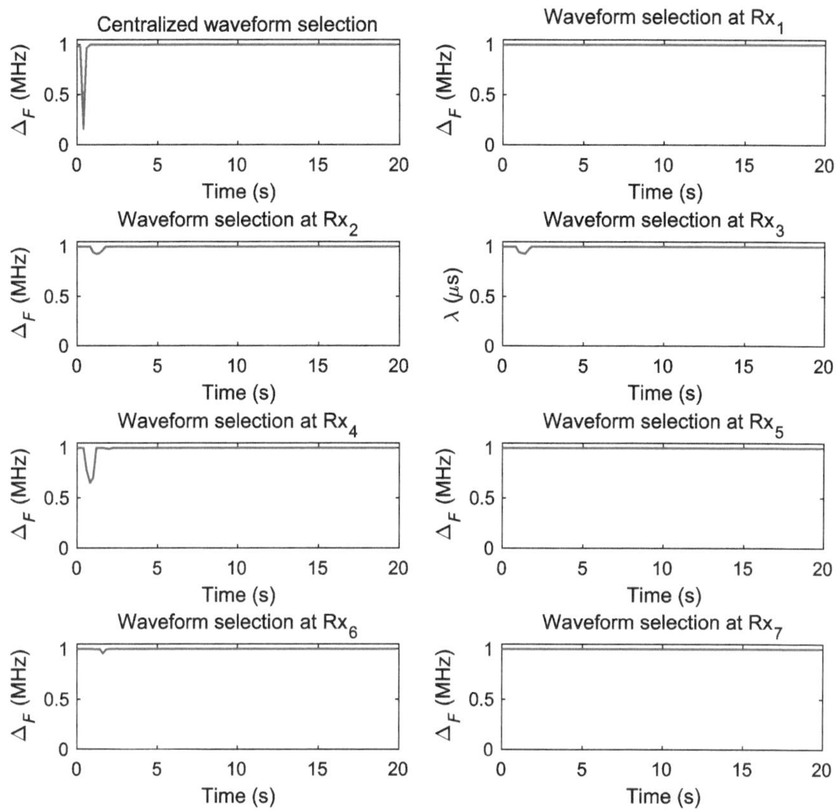

Figure 5.6 Patterns of selected waveform frequency sweeps.

performance degradation, the distributed processing can provide many important advantages over the centralized one as highlighted in Section 5.1. More importantly, the results in Fig. 5.3 also confirm the performance superiority of the adaptive waveform selection over the fixed waveform setting.

Fig. 5.4 shows the RMSE performance of the distributed target tracking and waveform selection algorithm when utilizing waveforms selected at different receivers (the patterns of selected waveform parameters are plotted in Figs. 5.5 and 5.6). We can observe that the RMSE remains almost identical regardless of which waveform is used. Therefore, the waveform selected by any receiver can be used at the transmitter without affecting the performance of the system. This endows the system with enhanced resilience and robustness.

5.5 Summary

This chapter has presented an algorithm which performs joint target tracking and waveform selection over the receivers of a multistatic radar system in a fully decentralized manner. In particular, the receivers form a connected network where they can communicate their measurements and track estimates to their immediate neighbors. Target tracking and waveform selection are performed at each receiver exploiting the information available in its closed neighborhood. Such a distributed processing strategy offers many important advantages over its centralized counterparts presented in Chapters 2–4 in terms of not only cost savings for the operation and maintenance of the system, but also increasing the robustness of the system to link or node failure.

PART 2

Optimal geometry analysis

Background and purpose

Relative sensor-target geometry is an important factor that can significantly impact the performance of target localization and tracking. Therefore, it is critical to determine the locations of the sensors (i.e., the locations of transmitters and receivers in the context of multistatic radar) such that the localization and tracking performance is optimized. Such an optimization problem is often referred to as *optimal sensor placement*. The literature of optimal sensor placement can be categorized into two different formulations: optimal control and parameter optimization.

The optimal control formulation is usually exploited for trajectory optimization, also known as optimal path planning, for moving sensor platforms (see e.g., [55–58]). The idea is to minimize the dynamic estimation error under the optimal control framework. The sensor trajectories can be decided via one-step or multiple-step look-ahead optimizations. Trajectory optimization is often subject to various nonlinear constraints including physical and geometric types such as minimum sensor-to-target clearance, maximum sensor-to-sensor distances to retain communication links, and obstacle and threat avoidance. Closed-form solutions typically do not exist for this optimal control formulation, and numerical approaches, for example, the parameterized unconstrained optimization [58], the gradient-descent method [55,56], and the differential inclusion solution [55], must be used.

The parameter optimization formulation, on the other hand, aims to determine the optimal sensor positions given a prior information about the target position by minimizing the uncertainty of the target position estimate (see e.g., [59–65]). This formulation is commonly known as optimal geometry analysis. In contrast to the optimal control formulation, the parameter optimization formulation, as is often the case, can be analytically solved. Such analytical solutions reveal important insights into how the sensor-target geometry affects the target estimation performance. The knowledge of analytical optimal geometries can also help to generate useful tactical strategies for optimizing the sensor path in order to increase fuel efficiency and sustain communication distance constraints while at the same time maintaining an optimal target estimation performance.

Optimal geometries have been analytically derived for various target localization problems such as the AOA-based localization in [59–61], the TOA-based localization [60,61], the TDOA-based localization in [62,63], the Doppler-based localization in [64,65], and the received signal strength-based localization in [61]. Most of these works characterize the target estimation performance based on the FIM which is the inverse of the CRLB. Specifically, the determinant of the FIM is utilized as the objective function to be maximized for geometry optimization. Maximizing the determinant of the FIM is in fact equivalent to minimizing the area of the estimation confidence area. This criterion is often referred to as the D-optimality criterion in the context of optimal experimental design [66]. Other criteria include the A-optimality criterion (i.e., minimizing the trace of the CRLB) and the E-optimality criterion (i.e., minimizing the maximum eigenvalue of the CRLB). In contrast to the A- and E-optimality criteria which are affected by linear transformations of the output and scale changes in the parameters, the D-optimality criterion is invariant under these transformations [66].

This part of the book presents optimal geometry analyses for the problem of target localization using TOA measurements collected by a multistatic radar system. The optimal geometry analyses are conducted based on the D-optimality criterion. Two different multistatic radar configurations are considered:

- Chapter 6: a single transmitter and multiple receivers,

- Chapter 7: multiple independent bistatic channels.

The accuracy of the analytical findings is verified via simulation examples with numerical solutions and sensor trajectory optimization.

CHAPTER 6

Optimal geometries for multistatic target localization with one transmitter and multiple receivers

Contents

6.1 Introduction and problem formulation

This chapter presents an optimal geometry analysis for the problem of 2D TOA-based target localization using a multistatic radar system with one transmitter and multiple receivers. In the literature, optimal geometries were derived in [67] for the MIMO radar configuration in which the number of transmitters and receivers is each greater than or equal to three. This work also showed that the estimation error increases by a factor of 2 if the optimal geometries derived for a configuration with M transmitters and N receivers is applied to a configuration with one transmitter and MN receivers. However, as discussed in [67,68], such geometries are not optimal for the case of one transmitter and multiple receivers because the number of transmitters is smaller than three in this case. An optimal geometry for one transmitter and multiple receivers was derived in [68] for the special case of an even number of receivers having identical noise variances. In contrast to [68], this chapter considers a general scenario where the number of receivers is arbitrary and the noise variances can vary across the receivers.

Fig. 6.1 depicts the multistatic TOA-based localization problem under consideration, where $\boldsymbol{p} = [p_x, p_y]^T$ denotes the target position to be estimated, $\boldsymbol{t} = [t_x, t_y]^T$ denotes the transmitter position, and $\boldsymbol{r}_i = [r_{x,i}, r_{y,i}]^T$,

Signal Processing for Multistatic Radar Systems
https://doi.org/10.1016/B978-0-12-815314-7.00015-9

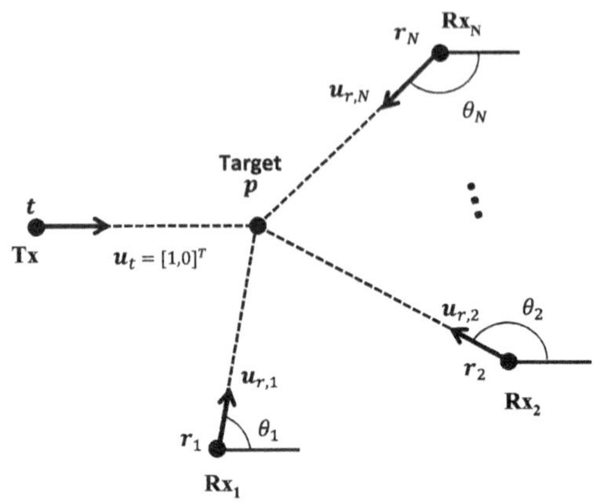

Figure 6.1 Localization geometry with one transmitter and multiple receivers.

$1 \le i \le N$, denotes the position of receiver i. With no loss of generality, the coordinate system is rotated so that the unit vector pointing from the transmitter to the target is $\boldsymbol{u}_t = [1, 0]^T$.

The bistatic TOA measurement τ_i taken at receiver i is related to the target position $\boldsymbol{p}$, the transmitter position $\boldsymbol{t}$ and the receiver position $\boldsymbol{r}_i$ through the nonlinear equation

$$\tilde{\tau}_i(\boldsymbol{p}) = \tau_i(\boldsymbol{p}) + e_i, \quad \tau_i(\boldsymbol{p}) = \frac{\|\boldsymbol{p} - \boldsymbol{t}\| + \|\boldsymbol{p} - \boldsymbol{r}_i\|}{c} \tag{6.1}$$

where c is the signal propagation speed, and e_i accounts for the measurement error which is modeled as an independent zero-mean Gaussian noise. To ensure the generality of our geometry analysis, we allow the error variances $E\{e_i^2\}$ to have different values for each receiver.

By multiplying (6.1) with the signal propagation speed c, we obtain a total bistatic range measurement equation

$$\tilde{d}_i(\boldsymbol{p}) = d_i(\boldsymbol{p}) + n_i, \quad d_i(\boldsymbol{p}) = \|\boldsymbol{p} - \boldsymbol{t}\| + \|\boldsymbol{p} - \boldsymbol{r}_i\|, \tag{6.2}$$

where $n_i = ce_i$ is the range error with $E\{n_i^2\} = \sigma_i^2$ (i.e., $E\{e_i^2\} = \sigma_i^2/c^2$). Stacking (6.2) for $i = 1, \ldots, N$ yields

$$\tilde{\boldsymbol{d}} = [\tilde{d}_1, \ldots, \tilde{d}_N]^T = \boldsymbol{d}(\boldsymbol{p}) + \boldsymbol{n} = [d_1, \ldots, d_N]^T + [n_1, \ldots, n_N]^T. \tag{6.3}$$

The covariance matrix for the error vector $\boldsymbol{n}$ is given by

$$\boldsymbol{\Sigma} = E\{\boldsymbol{n}\boldsymbol{n}^T\} = \mathrm{diag}(\sigma_1^2, \ldots, \sigma_N^2). \tag{6.4}$$

The objective of TOA localization is to estimate $\boldsymbol{p}$ from $\tilde{\boldsymbol{d}}$ via (6.3). Obtaining a unique solution for $\boldsymbol{p}$ requires at least three receivers ($N \geq 3$) since two TOA ellipses may intersect at two distinct points in the case of two receivers ($N = 2$), giving rise to a "ghost" target. Two receivers could still be sufficient if some prior information as regards the region of target position was available such that the solution ambiguity can be resolved. Solving (6.3) for $\boldsymbol{p}$ is a nonlinear estimation problem that was addressed in [68–70]. In this chapter, the algorithm used for target localization is assumed to be nearly efficient and unbiased so that the estimation error covariance can be approximated by the CRLB (i.e., the inverse of the FIM).

The FIM for the multistatic TOA localization problem under consideration is expressed by

$$\boldsymbol{\Phi} = \boldsymbol{J}_\circ^T \boldsymbol{\Sigma}^{-1} \boldsymbol{J}_\circ. \tag{6.5}$$

Here, $\boldsymbol{J}_\circ$ is the Jacobian matrix of $\boldsymbol{d}(\boldsymbol{p})$ with respect to $\boldsymbol{p}$:

$$\boldsymbol{J}_\circ = \begin{bmatrix} (\boldsymbol{u}_t + \boldsymbol{u}_{r,1})^T \\ (\boldsymbol{u}_t + \boldsymbol{u}_{r,2})^T \\ \vdots \\ (\boldsymbol{u}_t + \boldsymbol{u}_{r,N})^T \end{bmatrix} \tag{6.6}$$

where $\boldsymbol{u}_{r,i} = [\cos\theta_i, \sin\theta_i]^T$ with θ_i being the bearing angle of receiver i as depicted in Fig. 6.1. In fact, $\boldsymbol{u}_{r,i}$ is a unit vector pointing from $\boldsymbol{r}_i$ to $\boldsymbol{p}$. Using (6.4)–(6.6), we can re-express the FIM as

$$\begin{aligned} \boldsymbol{\Phi} &= \sum_{i=1}^{N} \frac{1}{\sigma_i^2} (\boldsymbol{u}_t + \boldsymbol{u}_{r,i})(\boldsymbol{u}_t + \boldsymbol{u}_{r,i})^T \\ &= \sum_{i=1}^{N} \frac{1}{\sigma_i^2} \begin{bmatrix} (1+\cos\theta_i)^2 & (1+\cos\theta_i)\sin\theta_i \\ (1+\cos\theta_i)\sin\theta_i & \sin^2\theta_i \end{bmatrix}. \end{aligned} \tag{6.7}$$

It is interesting to note that according to (6.7) the FIM does not depend on target-transmitter and target-receiver distances. For given measurement error variances $\sigma_1^2, \ldots, \sigma_N^2$, the FIM and thus the localization performance only depend on $\theta_1, \ldots, \theta_N$. Therefore, the geometry optimization problem boils down to finding the optimal angular separation between the transmitter and receivers with respect to the target. Specifically, we aim to maximize the determinant of the FIM with respect to $\theta_1, \ldots, \theta_N$:

$$\{\theta_1^*, \ldots, \theta_N^*\} = \underset{\{\theta_1,\ldots,\theta_N\}}{\arg\max} |\mathbf{\Phi}|, \tag{6.8}$$

which is the D-optimality criterion [66].

6.2 Optimal geometry analysis

After some algebraic manipulations, the determinant of the FIM in (6.7) can be expressed as

$$|\mathbf{\Phi}| = \sum_{i=1}^{N} \frac{(1+\cos\theta_i)^2}{\sigma_i^2} \sum_{i=1}^{N} \frac{\sin^2\theta_i}{\sigma_i^2} - \left(\sum_{i=1}^{N} \frac{(1+\cos\theta_i)\sin\theta_i}{\sigma_i^2}\right)^2. \tag{6.9}$$

The solution for maximizing $|\mathbf{\Phi}|$ over $\{\theta_1, \ldots, \theta_N\}$ is given by the following theorem.

Theorem 6.1. *The maximum determinant of* $\mathbf{\Phi}$ *is*

$$|\mathbf{\Phi}|_{\max} = \frac{27}{16}\left(\left(\sum_{i=1}^{N} \frac{1}{\sigma_i^2}\right)^2 - \left(\sum_{i=1}^{N} \frac{b_i^*}{\sigma_i^2}\right)^2\right), \tag{6.10}$$

which is attained when

$$|\theta_1^*| = \cdots = |\theta_N^*| = \pi/3, \tag{6.11}$$

$$\{b_1^*, \ldots, b_N^*\} = \underset{b_i \in \{1,-1\}, 1\le i\le N}{\arg\min} \left(\sum_{i=1}^{N} \frac{b_i}{\sigma_i^2}\right)^2. \tag{6.12}$$

Here $b_i = \text{sgn}(\theta_i)$ *and* $\text{sgn}(\cdot)$ *stands for the signum function.*

Fig. 6.2 illustrates an example of the optimal placement of the receivers relative to the target and the transmitter following the results of Theorem 6.1. In what follows, we present two lemmas from which the proof of Theorem 6.1 follows immediately.

Lemma 6.2 (The maximally optimal geometry). *If* $|\theta_1^*| = \cdots = |\theta_N^*| = \pi/3$ *and there exists a sequence* b_i^* *satisfying*

$$\left(\sum_{i=1}^{N} \frac{b_i^*}{\sigma_i^2}\right)^2 = 0, \tag{6.13}$$

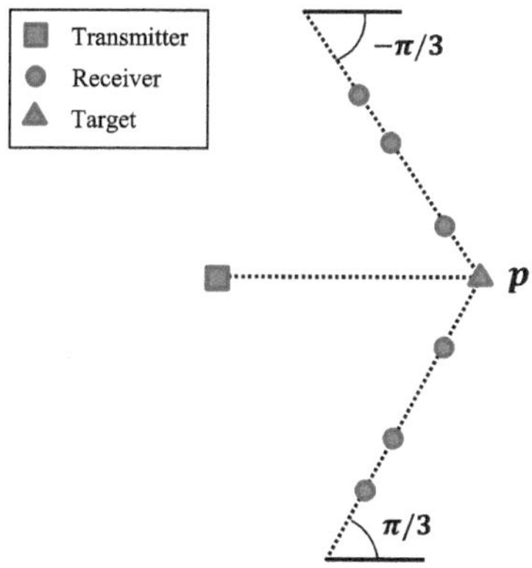

Figure 6.2 The optimal placement of the receivers relative to the target and the transmitter.

$|\mathbf{\Phi}|$ *in* (6.9) *attains its maximum value:*

$$|\mathbf{\Phi}|_{\max} = \frac{27}{16}\left(\sum_{i=1}^{N}\frac{1}{\sigma_i^2}\right)^2. \tag{6.14}$$

Proof. By letting $a_i = (1 + \cos\theta_i)$ with $0 \le a_i \le 2$ and hence $\sin^2\theta_i = a_i(2 - a_i)$, the first term in the right-hand side of (6.9) can be expressed as

$$\mathfrak{A} = \sum_{i=1}^{N}\frac{a_i^2}{\sigma_i^2}\sum_{i=1}^{N}\frac{a_i(2-a_i)}{\sigma_i^2} = \sum_{i=1}^{N}\frac{a_i^2}{\sigma_i^2}\left(2\sum_{i=1}^{N}\frac{a_i}{\sigma_i^2} - \sum_{i=1}^{N}\frac{a_i^2}{\sigma_i^2}\right). \tag{6.15}$$

The following power mean inequality holds for any real numbers $a_1, \ldots, a_N$ [71]:

$$\left(\sum_{i=1}^{N}\frac{a_i}{\sigma_i^2}\right)^2 \le \sum_{i=1}^{N}\frac{1}{\sigma_i^2}\sum_{i=1}^{N}\frac{a_i^2}{\sigma_i^2} \tag{6.16}$$

with equality if and only if $a_1 = \cdots = a_N$. As $a_i \ge 0$ for all $i = 1, \ldots, N$, we have

$$\sum_{i=1}^{N}\frac{a_i}{\sigma_i^2} \le \left(\sum_{i=1}^{N}\frac{1}{\sigma_i^2}\right)^{1/2}\left(\sum_{i=1}^{N}\frac{a_i^2}{\sigma_i^2}\right)^{1/2} \tag{6.17}$$

Substituting (6.17) into (6.15) yields

$$\mathfrak{A} \le 2\left(\sum_{i=1}^{N}\frac{1}{\sigma_i^2}\right)^{1/2} b^{3/2} - b^2 \tag{6.18}$$

with $b = \sum_{i=1}^{N}(a_i^2/\sigma_i^2)$. Since $b \geq 0$, the right-hand side of (6.18) attains its maximum value of $\frac{27}{16}\left(\sum_{i=1}^{N}(1/\sigma_i^2)\right)^2$ if $b = \frac{9}{4}\sum_{i=1}^{N}(1/\sigma_i^2)$. Therefore, we have

$$\mathfrak{A} \leq \frac{27}{16}\left(\sum_{i=1}^{N}\frac{1}{\sigma_i^2}\right)^2 \tag{6.19}$$

where the equality holds if

$$a_1 = \cdots = a_N \tag{6.20a}$$

and

$$b = \sum_{i=1}^{N}\frac{a_i^2}{\sigma_i^2} = \frac{9}{4}\sum_{1}^{N}\frac{1}{\sigma_i^2}. \tag{6.20b}$$

As a result, $\mathfrak{A}$ is maximized when $a_1 = \cdots = a_N = 3/2$ or $|\theta_1| = \cdots = |\theta_N| = \pi/3$.

Given $|\theta_1| = \cdots = |\theta_N| = \pi/3$, if a sequence b_i^* satisfying (6.13) exists, the second term in the right-hand side of (6.9) can be made zero by assigning the signs of θ_i according to b_i^*. This consequently maximizes $|\mathbf{\Phi}|$ at the value given in (6.14). □

Lemma 6.3 (The optimal geometry). *If the condition* (6.13) *cannot be met,* $|\mathbf{\Phi}|$ *in* (6.9) *is maximized when* $|\theta_1^*| = \cdots = |\theta_N^*| = \pi/3$ *and*

$$\{b_1^*, \ldots, b_N^*\} = \underset{b_i \in \{1,-1\}, 1 \leq i \leq N}{\arg\min} \left(\sum_{i=1}^{N}\frac{b_i}{\sigma_i^2}\right)^2 \tag{6.21}$$

with the maximum value given by (6.10).

Proof. If no sequence b_i satisfies (6.13), the second term in the right-hand side of (6.9) cannot be made zero for the case of $|\theta_1| = \cdots = |\theta_N| = \pi/3$. However, this term may still become zero with other angle arrangements rather than $|\theta_1| = \cdots = |\theta_N| = \pi/3$. In this case, maximizing the first term and minimizing the second term in the right-hand side of (6.9) does not provide a common solution.

It is important to note that the angular arrangements of $|\theta_1| = \cdots = |\theta_N| = \pi/3$ regardless of sign sequence are the critical points of $|\mathbf{\Phi}|$ even when the condition (6.13) is not satisfied (see Appendix A in Section 6.6 for the proof). Therefore, if other critical points do not exist, the solution

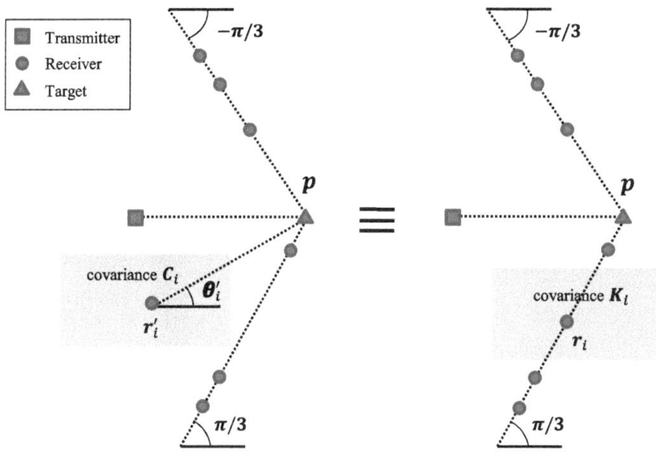

Figure 6.3 A relocation of receiver i from r_i ($\theta_i = \pi/3$) to r_i' ($\theta_i' \neq \pi/3$) is effectively equivalent to increasing the noise at r_i in the original optimal geometry from C_i to K_i, resulting in a reduction in the FIM determinant.

given in Lemma 6.3 will specify the optimal angular arrangement for the receivers. Appendix A does not directly rule out the existence of other critical points for $N \geq 3$. Next we will show that any rearrangement made to the optimal angular placement $\pm\pi/3$ rad of the receivers is equivalent to increasing the noise at the relocated receivers in the original optimal geometry. Therefore, Lemma 6.3 provides the optimal geometry solution regardless of whether or not there exist other critical points of $|\boldsymbol{\Phi}|$.

We begin with the optimal geometry given in Lemma 6.3 and note that

$$\left(\sum_{i=1}^{N} \frac{b_i^*}{\sigma_i^2}\right)^2 > 0. \tag{6.22}$$

Suppose that we change θ_i from $|\theta_i| = \pi/3$ (i.e., the optimal value given in Lemma 6.3) to a new value θ_i'. This corresponds to a relocation of receiver i from $\boldsymbol{r}_i$ to $\boldsymbol{r}_i'$ (see the left-hand side of Fig. 6.3 for an illustration). Our main interest is to see whether such an angular change would lead to an increase in the value of $|\boldsymbol{\Phi}|$ in the original optimal geometry.

The original contribution to the FIM of receiver i at position $\boldsymbol{r}_i$ can be expressed as

$$\boldsymbol{C}_i^{-1/2}\begin{bmatrix} 1+\cos\theta_i \\ \sin\theta_i \end{bmatrix}\begin{bmatrix} 1+\cos\theta_i & \sin\theta_i \end{bmatrix}\boldsymbol{C}_i^{-1/2}$$

$$
\begin{aligned}
&= \begin{bmatrix} (1+\cos\theta_i)/\sigma_i \\ \sin\theta_i/\sigma_i \end{bmatrix} \begin{bmatrix} (1+\cos\theta_i)/\sigma_i & \sin\theta_i/\sigma_i \end{bmatrix} \\
&= \frac{1}{\sigma_i^2} \begin{bmatrix} (1+\cos\theta_i)^2 & (1+\cos\theta_i)\sin\theta_i \\ (1+\cos\theta_i)\sin\theta_i & \sin^2\theta_i \end{bmatrix} \\
&= \frac{3}{4\sigma_i^2} \begin{bmatrix} 3 & b_i^*\sqrt{3} \\ b_i^*\sqrt{3} & 1 \end{bmatrix}
\end{aligned}
\tag{6.23}
$$

where $\boldsymbol{C}_i$ can be seen as the "covariance" matrix of TOA noise at receiver i:

$$
\boldsymbol{C}_i = \begin{bmatrix} \sigma_i^2 & 0 \\ 0 & \sigma_i^2 \end{bmatrix}. \tag{6.24}
$$

The expression in (6.23) is basically an outer product of $[1+\cos\theta_i, \sin\theta_i]^T$ with itself, weighted by the covariance matrix $\boldsymbol{C}_i$. In this sense, $\boldsymbol{C}_i$ represents the level of noise at receiver i. The noise levels are often compared based on the confidence region which is proportional to $|\boldsymbol{C}_i| = (\sigma_i^2)^2$ and the mean squared error which is equivalent to the average trace of $\boldsymbol{C}_i$, i.e., $\frac{1}{2}\operatorname{tr}\boldsymbol{C}_i = \sigma_i^2$. These two metrics will now be used in our analysis.

Consider the relocation of receiver i from $\boldsymbol{r}_i$ to $\boldsymbol{r}_i'$ by changing θ_i to θ_i' ($0 \le \theta_i' \le \pi$ if $b_i^* = 1$ or $-\pi \le \theta_i' \le 0$ if $b_i^* = -1$ to retain the sign of the bearing angle). The contribution of the relocated receiver to the FIM is [cf. (6.23)]

$$
\begin{aligned}
&\boldsymbol{C}_i^{-1/2} \begin{bmatrix} 1+\cos\theta_i' \\ \sin\theta_i' \end{bmatrix} \begin{bmatrix} 1+\cos\theta_i' & \sin\theta_i' \end{bmatrix} \boldsymbol{C}_i^{-1/2} \\
&\quad = \boldsymbol{K}_i^{-1/2} \begin{bmatrix} 1+\cos\theta_i \\ \sin\theta_i \end{bmatrix} \begin{bmatrix} 1+\cos\theta_i & \sin\theta_i \end{bmatrix} \boldsymbol{K}_i^{-1/2}.
\end{aligned}
\tag{6.25}
$$

Here, $\boldsymbol{K}_i$ is the equivalent covariance matrix of the TOA noise as if receiver i still remains at its original position $\boldsymbol{r}_i$ (see the right-hand side of Fig. 6.3 for an illustration). In fact, $\boldsymbol{K}_i$ can be explicitly written as

$$
\boldsymbol{K}_i = \begin{bmatrix} \sigma_{xi}^2 & 0 \\ 0 & \sigma_{yi}^2 \end{bmatrix} \tag{6.26}
$$

with

$$
\sigma_{xi} = \left(\frac{1+\cos\theta_i}{1+\cos\theta_i'}\right)\sigma_i = \frac{3}{2}\left(\frac{\sigma_i}{1+\cos\theta_i'}\right),
$$

$$\sigma_{yi} = \left(\frac{\sin\theta_i}{\sin\theta_i'}\right)\sigma_i = \frac{\sqrt{3}}{2}\left(\frac{\sigma_i}{|\sin\theta_i'|}\right).$$

The determinant of $\boldsymbol{K}_i$ is given by

$$|\boldsymbol{K}_i| = \sigma_{xi}^2\sigma_{yi}^2 = \frac{27}{16}\left(\frac{(\sigma_i^2)^2}{(1+\cos\theta_i')^2\sin^2\theta_i'}\right). \tag{6.27}$$

By noting that $(1+\cos\theta_i')^2\sin^2\theta_i' \leq 27/16$ where the equality is obtained if $|\theta_i'| = \pi/3$, we have

$$|\boldsymbol{K}_i| > |\boldsymbol{C}_i|. \tag{6.28}$$

This implies that the area of TOA error ellipse effectively increases if receiver i is relocated from its optimal angular position. The average trace of $\boldsymbol{K}_i$ is

$$\frac{1}{2}\operatorname{tr}\boldsymbol{K}_i = \frac{1}{2}(\sigma_{xi}^2 + \sigma_{yi}^2). \tag{6.29}$$

Using the inequality of arithmetic and geometric means [71], (6.27) and (6.28) yield

$$\frac{1}{2}\operatorname{tr}\boldsymbol{K}_i > \sigma_{xi}\sigma_{yi} > \frac{1}{2}\operatorname{tr}\boldsymbol{C}_i, \tag{6.30}$$

thus implying that the mean squared error also effectively increases if receiver i is relocated from its optimal angular position.

The above results indicate that the relocation of receiver i from $\boldsymbol{r}_i$ to $\boldsymbol{r}_i'$ is equivalent to keeping receiver i at its original optimal angular position but increasing its noise level. Any increase in receiver noise will reduce the determinant of the FIM and thus degrade the localization performance. For the optimal angular geometry described in Lemma 6.3, $|\boldsymbol{\Phi}|$ is shown in Section 6.6 (Appendix B) to be a decreasing function with respect to σ_i. Therefore, $|\boldsymbol{\Phi}|$ cannot be increased by relocating any receivers from its optimal angular position given in Lemma 6.3.

We have so far considered the case where θ_i' remains the same sign as θ_i. If θ_i' and θ_i have different signs, we can start with a geometry where $\theta_i = -b_i^*\pi/3$ and then follow the same noise injection arguments as above. We note that the geometry with $\theta_i = -b_i^*\pi/3$ is no longer optimal because the sign sequence $\{b_1, \ldots, b_N\}$ is non-optimal due to the sign flip (see (6.10)). As a result, the relocation of the receiver in this case even induce more noise into the system. Therefore, we rule out the possibility of $|\boldsymbol{\Phi}|$ exceeding its optimal value in this case. If the sign flip $\theta_i = -b_i^*\pi/3$ causes all angles

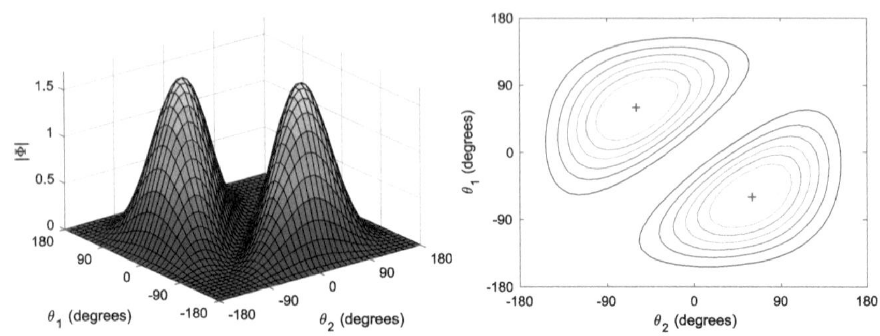

Figure 6.4 A plot of the FIM determinant for two receivers with respect to θ_1 and θ_2 with $\sigma_1 = 1$ and $\sigma_2 = 2$. Maxima occur at $\{\theta_1^*, \theta_2^*\} = \{60°, -60°\}$ and $\{60°, -60°\}$ indicated with '+' in the contour plot.

to have the same sign, $|\mathbf{\Phi}|$ becomes zero and this is equivalent to $\sigma_i \to \infty$. Thus, performing a relocation can effectively reduce the measurement noise at some receivers. However, this will never reduce the estimation covariance to the optimal covariance as shown above.

Although the relocation of one receiver is considered in the foregoing discussion, the generalization of the conclusions to multiple receivers is straightforward. As a result, we conclude that, if the condition (6.13) cannot be met, the receiver arrangement of $|\theta_1^*| = \cdots = |\theta_N^*| = \pi/3$ with $\text{sgn}(\theta_i^*)$ given by (6.21) maximizes $|\mathbf{\Phi}|$ and forms the optimal geometry. □

6.3 Examples

Example 1: Two receivers ($N = 2$)

The optimal angular geometries are

$$\{\theta_1^*, \theta_2^*\} = \{\pi/3, -\pi/3\}, \tag{6.31a}$$

$$\{\theta_1^*, \theta_2^*\} = \{-\pi/3, \pi/3\}, \tag{6.31b}$$

resulting in

$$|\mathbf{\Phi}|_{\max} = \frac{27}{4\sigma_1^2\sigma_2^2}. \tag{6.32}$$

Fig. 6.4 plots $|\mathbf{\Phi}|$ as a function of θ_1 and θ_2. We can observe that the maxima of $|\mathbf{\Phi}|$ match perfectly the derived optimal geometries.

Example 2: Three receivers ($N = 3$)

Given $\sigma_1 < \sigma_2 < \sigma_3$, the optimal angular geometries are

$$\{\theta_1^*, \theta_2^*, \theta_3^*\} = \{\pi/3, -\pi/3, -\pi/3\}, \tag{6.33a}$$

$$\{\theta_1^*, \theta_2^*, \theta_3^*\} = \{-\pi/3, \pi/3, \pi/3\}, \tag{6.33b}$$

thus resulting in

$$|\mathbf{\Phi}|_{\max} = \frac{27(\sigma_2^2 + \sigma_3^2)}{4\sigma_1^2\sigma_2^2\sigma_3^2}. \tag{6.34}$$

See Fig. 6.5 for example plots of $|\mathbf{\Phi}|$.

Example 3: Four receivers ($N = 4$)

Given $\sigma_1 < \sigma_2 < \sigma_3 < \sigma_4$, the optimal geometries are

$$\{\theta_1^*, \theta_2^*, \theta_3^*, \theta_4^*\} = \{\pi/3, -\pi/3, -\pi/3, -\pi/3\}, \tag{6.35a}$$

$$\{\theta_1^*, \theta_2^*, \theta_3^*, \theta_4^*\} = \{-\pi/3, \pi/3, \pi/3, \pi/3\}, \tag{6.35b}$$

if $1/\sigma_1^2 > (1/\sigma_2^2 + 1/\sigma_3^2)$. Otherwise, the optimal geometries are

$$\{\theta_1^*, \theta_2^*, \theta_3^*, \theta_4^*\} = \{\pi/3, -\pi/3, -\pi/3, \pi/3\}, \tag{6.36a}$$

$$\{\theta_1^*, \theta_2^*, \theta_3^*, \theta_4^*\} = \{-\pi/3, \pi/3, \pi/3, -\pi/3\}. \tag{6.36b}$$

Example 4: Even number of receivers with i.i.d. noise

If $\sigma_1 = \cdots = \sigma_N = \sigma$ and N is even, which is the case considered in [68], the maximally optimal angular geometries are

$$\theta_1^* = \cdots = \theta_{N/2}^* = \pi/3, \tag{6.37a}$$

$$\theta_{N/2+1}^* = \cdots = \theta_N^* = -\pi/3, \tag{6.37b}$$

yielding

$$|\mathbf{\Phi}|_{\max} = \frac{27\,N^2}{16\sigma^4}. \tag{6.38}$$

The optimality is not affected by the permutations of the receivers because different receivers have the same noise variance. The derived geometry here differs from that presented in [68] which has a half of the receivers allocated the bearing angle of 70.53° (i.e., 1.231 rad) and the remaining receivers allocated at the bearing angle of −70.53°. This difference can be explained by noting that [68] utilized the A-optimality criterion instead of the D-optimality criterion as in our analysis.

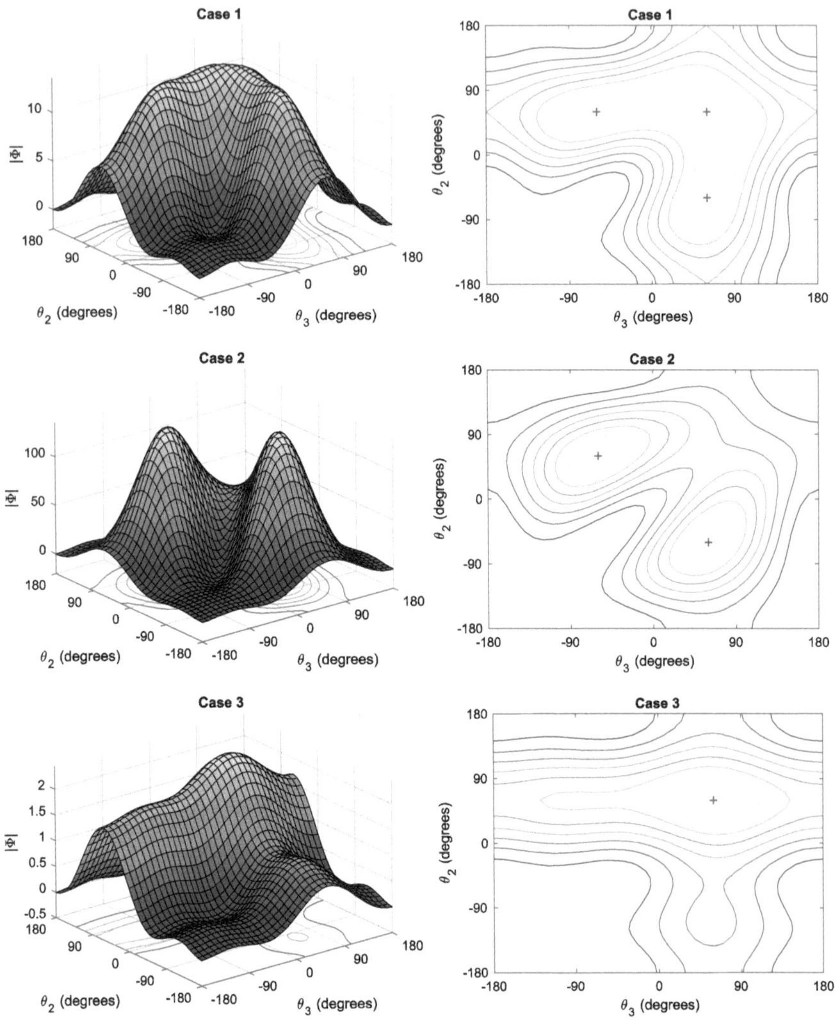

Figure 6.5 Plots of the FIM determinant for three receivers with respect to θ_2 and θ_3 for $\theta_1 = -60°$ (maxima indicated with '+'). Case 1: $\sigma_1 = \sigma_2 = \sigma_3 = 1$ (maxima occur at $\{\theta_2^*, \theta_3^*\} = \{60°, 60°\}$, $\{60°, -60°\}$ and $\{-60°, 60°\}$). Case 2: $\sigma_1 = 1, \sigma_2 = \sigma_3 = 0.5$ (maxima occur at $\{\theta_2^*, \theta_3^*\} = \{60°, -60°\}$ and $\{-60°, 60°\}$). Case 3: $\sigma_1 = 1, \sigma_2 = 2$ and $\sigma_3 = 3$ (maximum occurs at $\{\theta_2^*, \theta_3^*\} = \{60°, 60°\}$).

Example 5: Odd number of receivers with i.i.d. noise

If $\sigma_1 = \cdots = \sigma_N = \sigma$ and N is odd, we have

$$\theta_1^* = \cdots = \theta_{(N-1)/2}^* = \pi/3, \tag{6.39a}$$

$$\theta^*_{(N+1)/2} = \cdots = \theta^*_N = -\pi/3, \tag{6.39b}$$

and therefore

$$|\mathbf{\Phi}|_{\max} = \frac{27(N^2-1)}{16\sigma^4}. \tag{6.40}$$

The optimality is also not affected by the permutations of the receivers in this case. An illustration for this example can be found in Fig. 6.5 (case 1) for $N = 3$.

6.4 Simulations

6.4.1 Numerical solutions

To verify the derived optimal geometry solutions, the global maxima of $|\mathbf{\Phi}|$ in (6.8) are numerically calculated via the genetic algorithm [72]. The iterations of the genetic algorithm are stopped once the average relative difference in the value of best fitness function over 50 consecutive generations is smaller than 10^{-20} or when the algorithm reaches 1000 iterations. To ensure all solutions are captured in the case there exist multiple optimal geometries, we divide the search region into $[-180°, 0°]$ and $[0°, 180°]$ for each angle and execute the genetic algorithm over these regions separately.

The numerical solutions are obtained by averaging the results over 500 simulation runs after wrong solutions are eliminated and all different realizations of the same angular geometry are transformed into a common representative realization. Tables 6.1–6.3 report the numerical solutions for optimal geometries for $N = 2, 3$ and 4 in various noise scenarios. The numerical solutions are observed to perfectly match the derived optimal geometries presented in Section 6.3.

Table 6.1 Genetic algorithm solutions for $N = 2$.

	σ_1	σ_2	θ_1	θ_2
Case 1†	1	1	60.00°	−60.00°
Case 2†	1	2	60.00°	−60.00°
Case 3†	1	3	60.00°	−60.00°

† Matching the analytical result in Example 1.

Table 6.2 Genetic algorithm solutions for $N = 3$.

	σ_1	σ_2	σ_3	θ_1	θ_2	θ_3
Case 1[a,†]	1	2	3	60.00°	−60.00°	−60.00°
Case 2[b,†]	1	$2/\sqrt{3}$	2	60.00°	−60.00°	−60.00°

[a] The condition (6.13) is not satisfied.
[b] The condition (6.13) is satisfied.
[†] Matching the analytical result in Example 2.

Table 6.3 Genetic algorithm solutions for $N = 4$.

	θ_1	θ_2	θ_3	θ_4
Case 1[a]	60.00°	60.00°	−60.00°	−60.00°
Case 2[b]	60.00°	−60.00°	−60.00°	−60.00°
Case 3[c]	60.00°	−60.00°	−60.00°	60.00°

[a] Case 1 with $\sigma_1 = \sigma_2 = \sigma_3 = \sigma_4 = 1$. The permutations of the receivers are allowed as the error variances are equal (matching the analytical result in Example 4).
[b] Case 2 with $\sigma_1 = 1$, $\sigma_2 = 2$, $\sigma_3 = 3$, $\sigma_4 = 4$ (matching the analytical result in the first case of Example 3).
[c] Case 3 with $\sigma_1 = 1.8$, $\sigma_2 = 2$, $\sigma_3 = 3$, $\sigma_4 = 4$ (matching the analytical result in the second case of Example 3).

6.4.2 Sensor trajectory optimization

Trajectory optimization for moving sensor platforms is effectively another form of optimal sensor placement. In this section, we present simulation studies on trajectory optimization for unmanned aerial vehicle (UAV) platforms, where UAVs act as moving receiver platforms. In the simulations, the transmitter is fixed at $\boldsymbol{t} = [-20, 0]^T$ km while the target position is set to $\boldsymbol{p} = [0, 0]^T$ km. Our main interest is in comparing the analytical optimal geometries given in Section 6.2 to the final geometries formed by the UAVs.

The UAV trajectories are optimized and controlled based on maximizing the determinant of the FIM via gradient-descent optimization [56]:

$$\boldsymbol{s}(n+1) = \boldsymbol{s}(n) + \boldsymbol{M}(n)\,\frac{\partial J_T(\boldsymbol{s}(n))}{\partial \boldsymbol{s}(n)} \tag{6.41}$$

where $\boldsymbol{s}(n)$ is a vector of UAV positions at time instant n, $\boldsymbol{M}(n)$ is a time-varying step-size matrix that forces each UAV to follow the direction guided by the gradient and maintain a uniform distance between successive time instants, and $J_T(\boldsymbol{s}(n)) = |\boldsymbol{\Phi}|$ is the objective function to be maximized. In the simulations, we employ the first-order finite difference approximation [73] to calculate the gradient of $J_T(\boldsymbol{s}(n))$ where the

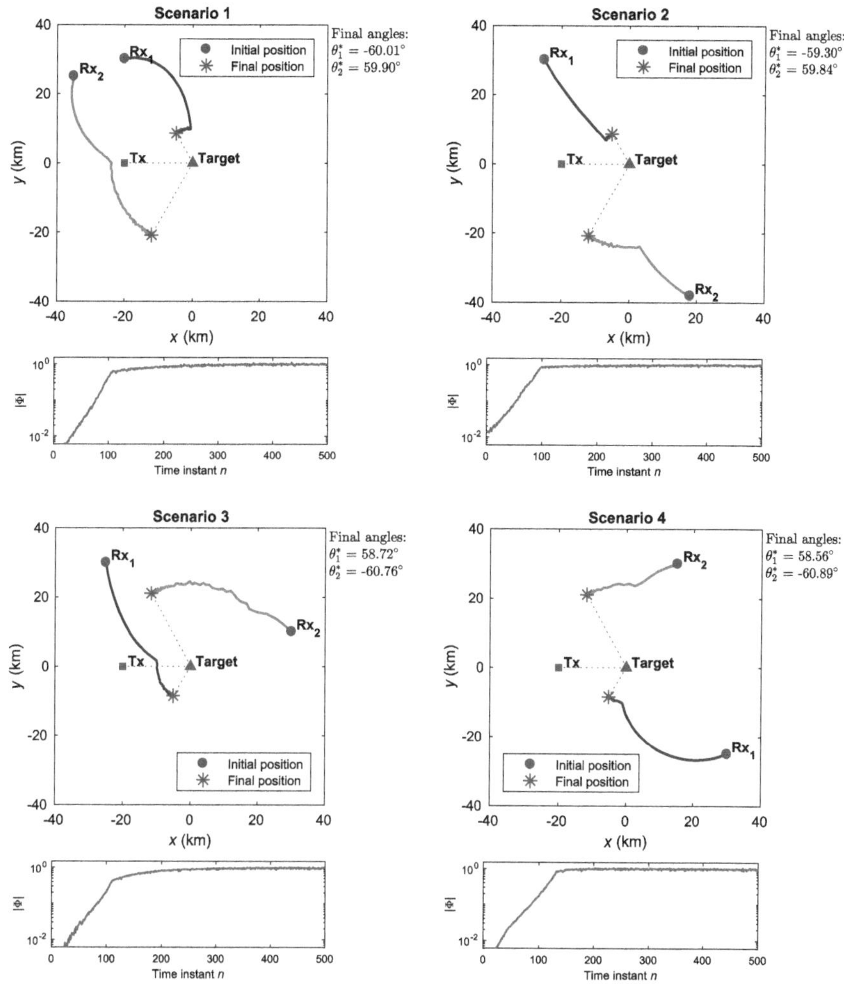

Figure 6.6 Trajectory optimization for two UAVs with the hard constraints of $l_1 = 10$ km and $l_2 = 24$ km. The final geometries formed by the UAVs closely match the analytical geometry solution in Example 1.

constant in (28) of [56] is set to $\Delta = 100$ m. Note that, since the true position of the target is unknown in practical scenarios, $|\mathbf{\Phi}|$ is computed from the estimated target position calculated using the iterative linearized least-squares estimator [74]. The variance of TOA error at a receiver is modeled as range-dependent [75], using $E\{e^2\} = \varrho_\circ^2(R_T^2 R_R^2)/R_\circ^4$, where R_T and R_R denote the distances from the target to the transmitter and the receiver, respectively. Here, we set the reference range $R_\circ$ to 20 km, and

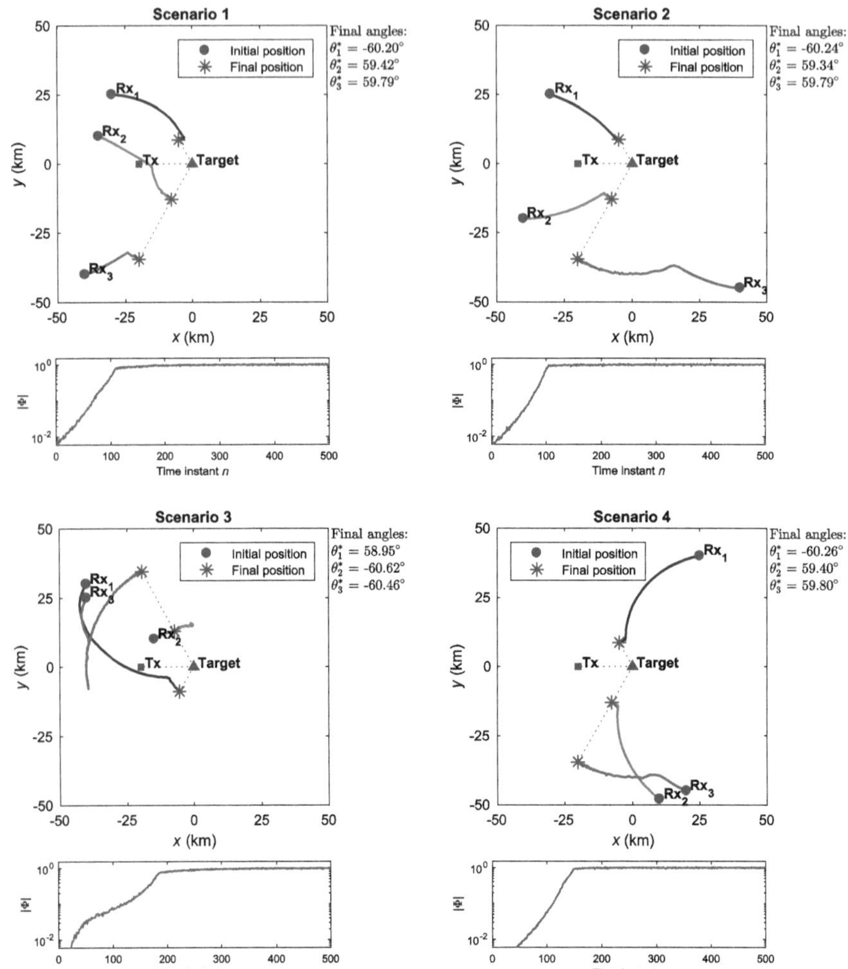

Figure 6.7 Trajectory optimization for three UAVs with the hard constraints of $l_1 = 10$ km, $l_2 = 15$ km and $l_3 = 40$ km (i.e., $\sigma_1 < \sigma_2 < \sigma_3$ when all UAVs reach their hard constraints). The final geometries formed by the UAVs closely match the analytical geometry solution in Example 2.

the constant $\varrho_\circ$ to 1 μs. The time step is $\Delta_T = 10$ s, and the maximum speed of UAVs is 30 m/s. Thus, the maximum distance which a UAV can travel in one time step is 300 m. The speed of signal propagation is set to $c = 3 \times 10^8$ m/s. A hard constraint is imposed on the UAVs to force them to maintain a minimum clearance (denoted as l_i for receiver i) from the target.

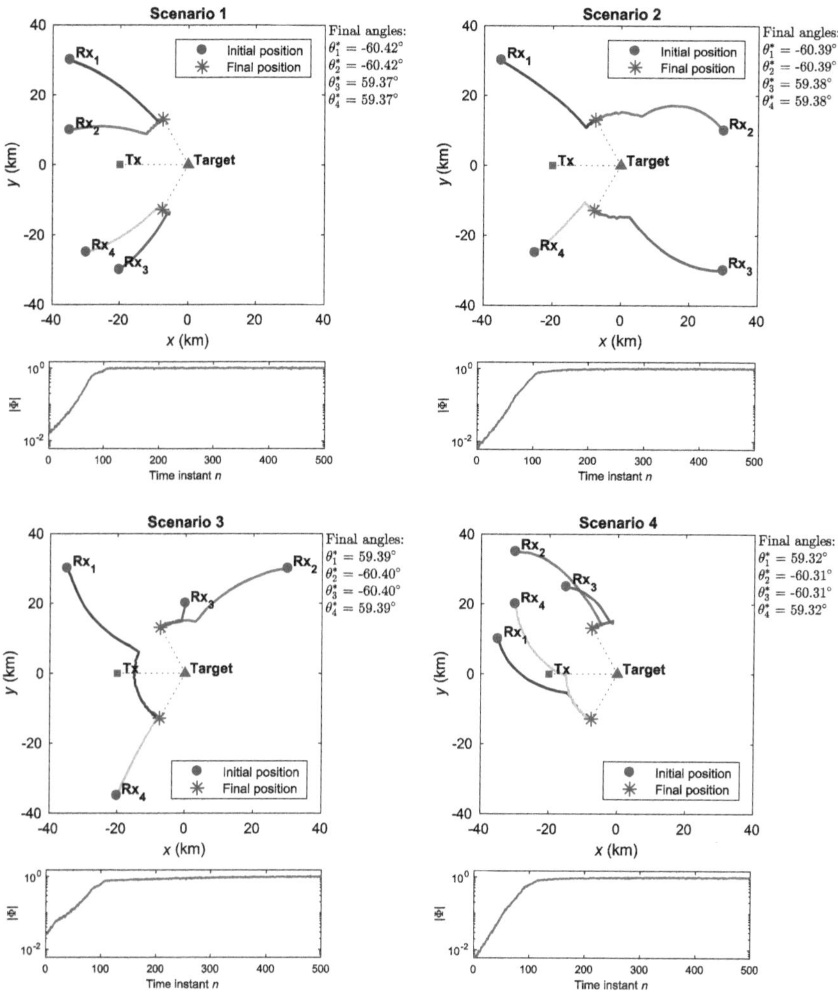

Figure 6.8 Trajectory optimization for four UAVs with the hard constraints of $l_1 = l_2 = l_3 = l_4 = 15$ km (i.e., identical error variances when all UAVs reach their hard constraints). The final geometries formed by the UAVs closely match the analytical geometry solution in Example 4.

We consider three examples with $N = 2, 3$ and 4 UAVs for various scenarios with different initial positions. The simulated optimal UAV trajectories are shown in Figs. 6.6–6.9. We observe that the UAVs tend to move towards the target until they reach their minimum allowed distances from the target because the TOA error variances are reduced as the UAVs get closer to the target. After reaching the minimum distance constraints,

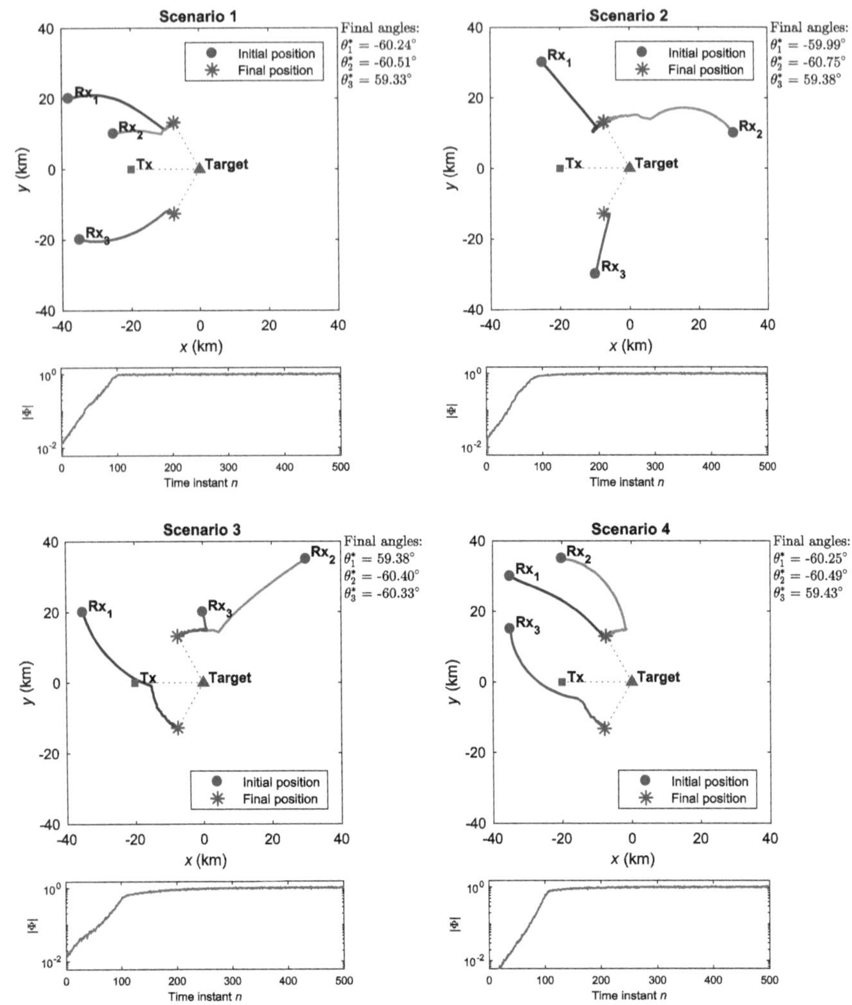

Figure 6.9 Trajectory optimization for three UAVs with the hard constraints of $l_1 = l_2 = l_3 = 15$ km (i.e., identical error variances when all UAVs reach their hard constraints). The final geometries formed by the UAVs closely match the analytical geometry solution in Example 5.

the UAVs move around these circular constraints until they form an optimal angular geometry. If multiple optimal geometries exist, the final angular configuration formed by the UAVs is influenced by the initial UAV positions. In Figs. 6.6–6.9, the final UAV geometries are observed to closely coincide with the optimal geometries derived in Section 6.2.

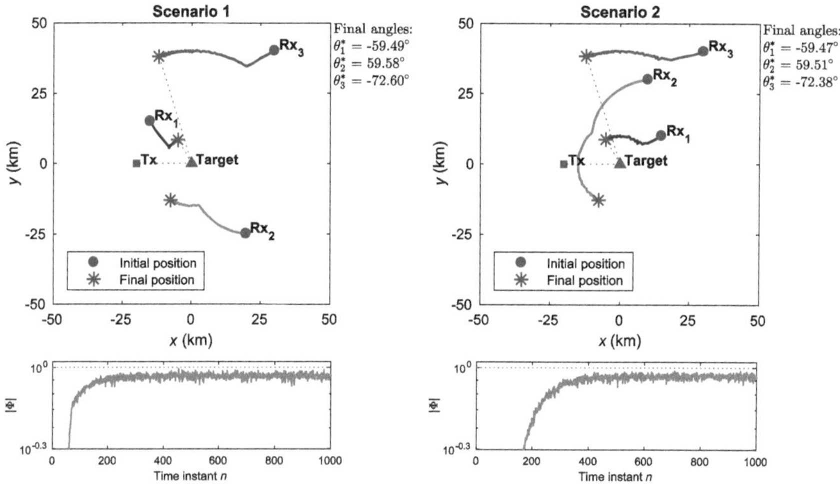

Figure 6.10 Illustrations of the UAVs forming a suboptimal angular configuration. The hard constraints are $l_1 = 10$ km, $l_2 = 15$ km and $l_3 = 40$ km. In these scenarios, $\{\theta_1, \theta_2, \theta_3\} = \{-60°, 60°, -60°\}$ is a saddle point of $|\mathbf{\Phi}|$ with the gradient of $|\mathbf{\Phi}|$ approximately zero for $-73° < \theta_3 < -50°$, and $\theta_1 \approx -60°$ and $\theta_2 \approx 60°$, causing the UAVs to get stuck in a suboptimal angular separation.

Figs. 6.6–6.9 also plot the evolution of the FIM determinant after normalized by the optimal FIM determinant value derived in Section 6.2. The normalized determinant of the FIM is observed to increase over time and converge to 1. This verifies that the final UAV formations are indeed optimal. Note that there exist small differences between the simulation and analytical results because the FIM determinant is calculated using the target position estimate obtained from previous time step and the first-order finite difference method is used to approximate the gradient of the FIM determinant.

The UAVs may be trapped in a suboptimal saddle or local maximum point of the FIM determinant as demonstrated in Fig. 6.10. In such a situation, the analytical optimal geometries can be used to direct the UAVs to the correct optimal geometry configuration.

6.5 Summary

This chapter has presented an optimal geometry analysis for the TOA localization problem using a multistatic system with one transmitter and multiple receivers. The optimal geometries were analytically derived based on maximizing the determinant of the FIM which effectively minimizes the area of the estimation confidence region. Regardless of noise variances and number of receivers, the derived optimal geometries have a similar geometric configuration, requiring an angular separation of $\pm 60°$ between the transmitter and each receiver with respect to the target. A full characterization of the optimal geometries is given by Theorem 6.1. Simulation studies on numerical solutions using the genetic algorithm and UAV trajectory optimization were presented to verify the derived optimal geometries, where good agreement between the analytical and numerical results was observed.

6.6 Appendices

Appendix A

The determinant of the FIM in (6.9) can be rewritten as

$$|\mathbf{\Phi}| = \sum_{i=1}^{N} \sum_{j=1}^{N} \frac{\left(\sin\theta_i - \sin\theta_j + \sin(\theta_i - \theta_j)\right)^2}{2\sigma_i^2\sigma_j^2} \tag{6.42}$$

with its partial derivative given by

$$\frac{\partial|\mathbf{\Phi}|}{\partial\theta_i} = 2 \sum_{\substack{1\le j\le N \\ j\neq i}} \frac{\mathfrak{F}(\theta_i, \theta_j)}{\sigma_i^2\sigma_j^2} \tag{6.43}$$

where

$$\mathfrak{F}(\theta_i, \theta_j) = \left(\sin\theta_i - \sin\theta_j + \sin(\theta_i - \theta_j)\right)\left(\cos\theta_i + \cos(\theta_i - \theta_j)\right). \tag{6.44}$$

We observe that $\mathfrak{F}(\theta_i, \theta_j) = \mathfrak{F}(\theta_j, \theta_i) = 0$ when $\theta_i = \theta_j$, $\{\theta_i, \theta_j\} = \{\pi/3, -\pi/3\}$, or $\{\theta_i, \theta_j\} = \{-\pi/3, \pi/3\}$.

For $N = 2$, the critical points of $|\mathbf{\Phi}|$ are $\theta_1 = \theta_2$, $\{\theta_1, \theta_2\} = \{\pi/3, -\pi/3\}$ and $\{\theta_1, \theta_2\} = \{-\pi/3, \pi/3\}$ because $\mathfrak{F}(\theta_1, \theta_2) = \mathfrak{F}(\theta_2, \theta_1) = 0$ at these points (i.e., $\partial|\mathbf{\Phi}|/\partial\theta_1 = \partial|\mathbf{\Phi}|/\partial\theta_2 = 0$). We have $|\mathbf{\Phi}| = 0$ for $\theta_1 = \theta_2$, hence forming the worst geometry. In contrast, $|\mathbf{\Phi}| = 27/(4\sigma_1^2\sigma_2^2)$ for $\{\theta_1, \theta_2\} = \{\pi/3, -\pi/3\}$, or $\{\theta_1, \theta_2\} = \{-\pi/3, \pi/3\}$, which implies optimal angular geometries for $N = 2$.

For $N \geq 3$, we have $\partial|\mathbf{\Phi}|/\partial\theta_i = 0$ for all $i = 1, \ldots, N$, if $\mathfrak{F}(\theta_i, \theta_j) = \mathfrak{F}(\theta_j, \theta_i) = 0$ for all $i, j = 1, \ldots, N$. Correspondingly, the critical points of $|\mathbf{\Phi}|$ are given by the following angular configurations:

- $\theta_1 = \cdots = \theta_N$ (i.e., $|\mathbf{\Phi}| = 0$, leading to the worst geometry),

- $|\theta_i| = \pi/3$ for all $i = 1, \ldots, N$.

However, the existence of other critical points is not necessarily ruled out by (6.43) if $\mathfrak{F}(\theta_i, \theta_j) \neq 0$.

Appendix B

The partial derivative $\partial|\mathbf{\Phi}|/\partial\sigma_i$ is given by

$$\frac{\partial|\mathbf{\Phi}|}{\partial\sigma_i} = \frac{4(1+\cos\theta_i)\sin\theta_i}{\sigma_i^3}\sum_{j=1}^{N}\frac{(1+\cos\theta_j)\sin\theta_j}{\sigma_j^2} - \frac{2(1+\cos\theta_i)^2}{\sigma_i^3}\sum_{j=1}^{N}\frac{\sin^2\theta_j}{\sigma_j^2} - \frac{2\sin^2\theta_i}{\sigma_i^3}\sum_{j=1}^{N}\frac{(1+\cos\theta_j)^2}{\sigma_j^2}. \tag{6.45}$$

Given the optimal geometry with $|\theta_1^*| = \cdots = |\theta_N^*| = \pi/3$ and sign sequence b_i^* described in Lemma 6.3, we have

$$\frac{\partial|\mathbf{\Phi}|}{\partial\sigma_i} = -\frac{27}{4\sigma_i^3}\left\{\sum_{j=1}^{N}\frac{1}{\sigma_j^2} - b_i^*\sum_{j=1}^{N}\frac{b_j^*}{\sigma_j^2}\right\} \tag{6.46a}$$

$$= \begin{cases} -\dfrac{27}{2\sigma_i^3}\displaystyle\sum_{\{j|b_j^*=-1\}}\frac{1}{\sigma_j^2}, & \text{for } b_i^* = 1, \\ -\dfrac{27}{2\sigma_i^3}\displaystyle\sum_{\{j|b_j^*=1\}}\frac{1}{\sigma_j^2}, & \text{for } b_i^* = -1. \end{cases} \tag{6.46b}$$

Since the optimal geometries do not include the case of $b_1^* = \cdots = b_N^* = 1$ or $b_1^* = \cdots = b_N^* = -1$ for which $|\mathbf{\Phi}| = 0$, the case of $\partial|\mathbf{\Phi}|/\partial\sigma_i = 0$ is eliminated. Therefore, from (6.46b) we have

$$\frac{\partial|\mathbf{\Phi}|}{\partial\sigma_i} < 0 \text{ for all } \sigma_i, i = 1, \ldots, N. \tag{6.47}$$

As a result, $|\mathbf{\Phi}|$ is a decreasing function of σ_i given the optimal geometry in Lemma 6.3.

CHAPTER 7

Optimal geometries for multistatic target localization by independent bistatic channels

Contents

7.1 Introduction and problem formulation

This chapter presents an optimal geometry analysis for the 2D TOA-based target localization problem with a multistatic radar system consisting of multiple independent bistatic channels. This configuration of multistatic radar can be formed by two or more bistatic transmitter-receiver pairs of different type (e.g., monopulse or continuous-wave) operating in different frequency bands (e.g., L-, S-, C-, X-bands, etc.).

Fig. 7.1 depicts the geometry for the localization problem under consideration with N independent bistatic channels Tx_i–Rx_i $(i = 1, 2, ..., N)$. Here, $\boldsymbol{p} = [p_x, p_y]^T$ denotes the unknown target location, and $\boldsymbol{t}_i = [t_{x,i}, t_{y,i}]^T$ and $\boldsymbol{r}_i = [r_{x,i}, r_{y,i}]^T$ denote the transmitter and receiver positions of the ith bistatic channel, respectively. The TOA measurement τ_i of the ith bistatic channel is given by

$$\tilde{\tau}_i(\boldsymbol{p}) = \tau_i(\boldsymbol{p}) + e_i, \quad \text{with } \tau_i(\boldsymbol{p}) = \frac{\|\boldsymbol{p} - \boldsymbol{t}_i\| + \|\boldsymbol{p} - \boldsymbol{r}_i\|}{c} \tag{7.1}$$

where c is the speed of signal propagation and e_i is an independent Gaussian noise with zero mean and variance $E\{e_i^2\}$ accounting for measurement errors. The total bistatic range measurement $\tilde{d}_i$ at the ith channel can therefore be expressed as

$$\tilde{d}_i(\boldsymbol{p}) = d_i(\boldsymbol{p}) + n_i, \quad \text{with } d_i(\boldsymbol{p}) = \|\boldsymbol{p} - \boldsymbol{t}_i\| + \|\boldsymbol{p} - \boldsymbol{r}_i\|. \tag{7.2}$$

Signal Processing for Multistatic Radar Systems
https://doi.org/10.1016/B978-0-12-815314-7.00016-0

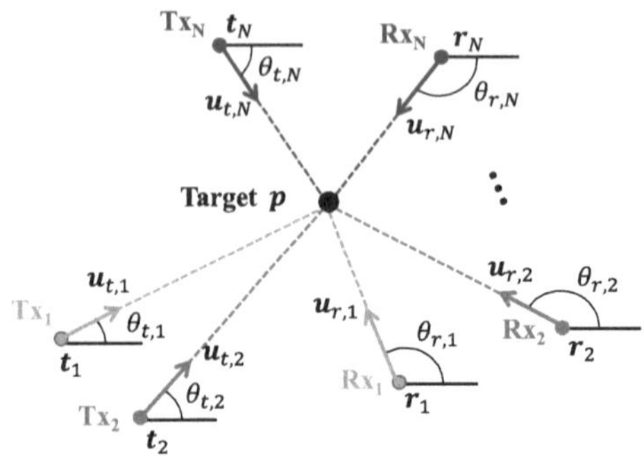

Figure 7.1 Multistatic TOA localization with multiple independent bistatic channels.

Here, $n_i = ce_i$ denotes the range measurement error with $E\{n_i^2\} = \sigma_i^2$ (i.e., $E\{e_i^2\} = \sigma_i^2/c^2$). Stacking (7.2) for $i = 1, 2, \ldots, N$ yields

$$\tilde{\boldsymbol{d}} = \boldsymbol{d}(\boldsymbol{p}) + \boldsymbol{n} = [d_1, d_2, \ldots, d_N]^T + [n_1, n_2, \ldots, n_N]^T \tag{7.3}$$

with

$$\boldsymbol{\Sigma} = E\{\boldsymbol{n}\boldsymbol{n}^T\} = \mathrm{diag}(\sigma_1^2, \sigma_2^2, \ldots, \sigma_N^2). \tag{7.4}$$

The objective of TOA localization is to estimate $\boldsymbol{p}$ from $\tilde{\boldsymbol{d}}$ using (7.3). Various solutions for multistatic TOA localization can be found in [68–70]. To ensure a unique solution for $\boldsymbol{p}$, at least three bistatic channels ($N \geq 3$) are required since two TOA ellipses may intersect at two distinct points in the case of two bistatic channels ($N = 2$). However, the solution ambiguity may be resolved given some prior knowledge about the region where the target is located, and thus two bistatic channels would be sufficient in this case.

The FIM for this multistatic TOA localization problem is given by

$$\boldsymbol{\Phi} = \boldsymbol{J}_\circ^T \boldsymbol{\Sigma}^{-1} \boldsymbol{J}_\circ \tag{7.5}$$

where $\boldsymbol{J}_\circ$ is the Jacobian matrix of $\boldsymbol{d}(\boldsymbol{p})$ with respect to $\boldsymbol{p}$ evaluated at the true value of $\boldsymbol{p}$:

$$\boldsymbol{J}_\circ = \begin{bmatrix} (\boldsymbol{u}_{t,1} + \boldsymbol{u}_{r,1})^T \\ (\boldsymbol{u}_{t,2} + \boldsymbol{u}_{r,2})^T \\ \vdots \\ (\boldsymbol{u}_{t,N} + \boldsymbol{u}_{r,N})^T \end{bmatrix}, \quad \boldsymbol{u}_{t,i} = \begin{bmatrix} \cos\theta_{t,i} \\ \sin\theta_{t,i} \end{bmatrix}, \quad \boldsymbol{u}_{r,i} = \begin{bmatrix} \cos\theta_{r,i} \\ \sin\theta_{r,i} \end{bmatrix}. \tag{7.6}$$

Note that $\boldsymbol{u}_{t,i}$ and $\boldsymbol{u}_{r,i}$ are the transmitter-to-target and receiver-to-target unit vectors as shown in Fig. 7.1. After some algebraic manipulations, we obtain

$$\boldsymbol{\Phi} = \sum_{i=1}^{N} \frac{1}{\sigma_i^2}(\boldsymbol{u}_{t,i} + \boldsymbol{u}_{r,i})(\boldsymbol{u}_{t,i} + \boldsymbol{u}_{r,i})^T. \tag{7.7}$$

Similar to Chapter 6, the localization algorithm under consideration is assumed to be nearly efficient and unbiased so that the estimation error covariance can be approximated by the CRLB, and the D-optimality criterion, i.e., maximizing the determinant of the FIM, is adopted for our geometry analysis. Noting that the FIM does not directly depend on the target-transmitter and target-receiver distances, we maximize the determinant of the FIM with respect to the angular separation between the transmitters and receivers:

$$\{\theta_{t,1}^*, \theta_{r,1}^*, \theta_{t,2}^*, \theta_{r,2}^*, \ldots, \theta_{t,N}^*, \theta_{r,N}^*\} = \underset{\{\theta_{t,1},\theta_{r,1},\theta_{t,2},\theta_{r,2},\ldots,\theta_{t,N},\theta_{r,N}\}}{\arg\max} |\boldsymbol{\Phi}|. \tag{7.8}$$

Observation 7.1. *The FIM remains unchanged by reflecting both the transmitter and receiver of a bistatic channel about the target (i.e., moving Tx_i from $\boldsymbol{t}_i$ to $2\boldsymbol{p} - \boldsymbol{t}_i$ and Rx_i from $\boldsymbol{r}_i$ to $2\boldsymbol{p} - \boldsymbol{r}_i$).*

Proof. Substituting $\boldsymbol{t}_i$ for $2\boldsymbol{p} - \boldsymbol{t}_i$ and $\boldsymbol{r}_i$ for $2\boldsymbol{p} - \boldsymbol{r}_i$ in (7.7), $(\boldsymbol{u}_{t,i} + \boldsymbol{u}_{r,i})$ becomes $-(\boldsymbol{u}_{t,i} + \boldsymbol{u}_{r,i})$, which does not change the FIM. □

Based on this observation, new optimal geometries can be generated from a given optimal geometry by reflecting one or more bistatic channels about the target.

7.2 Optimal geometry analysis

The FIM can be re-expressed as

$$\boldsymbol{\Phi} = \begin{bmatrix} \phi_{11} & \phi_{12} \\ \phi_{21} & \phi_{22} \end{bmatrix} \tag{7.9}$$

where

$$\phi_{11} = 2\sum_{i=1}^{N} \frac{1}{\sigma_i^2} \left(1 + \cos(\theta_{t,i} + \theta_{r,i})\right) \cos^2\left(\frac{\theta_{t,i} - \theta_{r,i}}{2}\right), \tag{7.10a}$$

$$\phi_{22} = 2\sum_{i=1}^{N} \frac{1}{\sigma_i^2} \left(1 - \cos(\theta_{t,i} + \theta_{r,i})\right) \cos^2\left(\frac{\theta_{t,i} - \theta_{r,i}}{2}\right), \tag{7.10b}$$

$$\phi_{12} = \phi_{21} = 2\sum_{i=1}^{N} \frac{1}{\sigma_i^2} \sin(\theta_{t,i} + \theta_{r,i}) \cos^2\left(\frac{\theta_{t,i} - \theta_{r,i}}{2}\right). \tag{7.10c}$$

Thus, the determinant of the FIM becomes

$$\begin{aligned} |\mathbf{\Phi}| = 4 \Bigg\{ &\left(\sum_{i=1}^{N} \frac{1}{\sigma_i^2} \cos^2\left(\frac{\theta_{t,i} - \theta_{r,i}}{2}\right)\right)^2 \\ &- \left(\sum_{i=1}^{N} \frac{1}{\sigma_i^2} \cos^2\left(\frac{\theta_{t,i} - \theta_{r,i}}{2}\right) \cos(\theta_{t,i} + \theta_{r,i})\right)^2 \\ &- \left(\sum_{i=1}^{N} \frac{1}{\sigma_i^2} \cos^2\left(\frac{\theta_{t,i} - \theta_{r,i}}{2}\right) \sin(\theta_{t,i} + \theta_{r,i})\right)^2 \Bigg\}. \end{aligned} \tag{7.11}$$

Using

$$\left(\sum_{i=1}^{N} a_i\right)^2 = \sum_{i=1}^{N} a_i^2 + 2\sum_{\{i,j\}\in C} a_i a_j \tag{7.12}$$

and the parameter transformation

$$\alpha_i = \frac{\theta_{t,i} - \theta_{r,i}}{2}, \quad \beta_i = \frac{\theta_{t,i} + \theta_{r,i}}{2}, \tag{7.13}$$

we obtain

$$|\mathbf{\Phi}| = 16 \sum_{\{i,j\}\in C} \frac{1}{\sigma_i^2 \sigma_j^2} \cos^2(\alpha_i) \cos^2(\alpha_j) \sin^2(\beta_i - \beta_j) \tag{7.14}$$

where C is a set of all 2-combinations of $\{1, 2, \ldots, N\}$. There are a total of $N!/(2!(N-2)!)$ combinations of $\{i, j\}$.

Since α_i and β_i are independent variables, the following inequality holds for any particular values of $\beta_1, \beta_2, \ldots, \beta_N$:

$$|\mathbf{\Phi}| \leq 16 \sum_{\{i,j\}\in C} \frac{1}{\sigma_i^2 \sigma_j^2} \sin^2(\beta_i - \beta_j) \tag{7.15}$$

with the equality satisfied when $\cos^2(\alpha_i) = 1$ for all $i = 1, 2, \ldots, N$ (i.e., $\alpha_i = 0$ or $\alpha_i = \pi$; hence $\theta_{t,i} = \theta_{r,i} = \beta_i$). Thus, we have the following formal result.

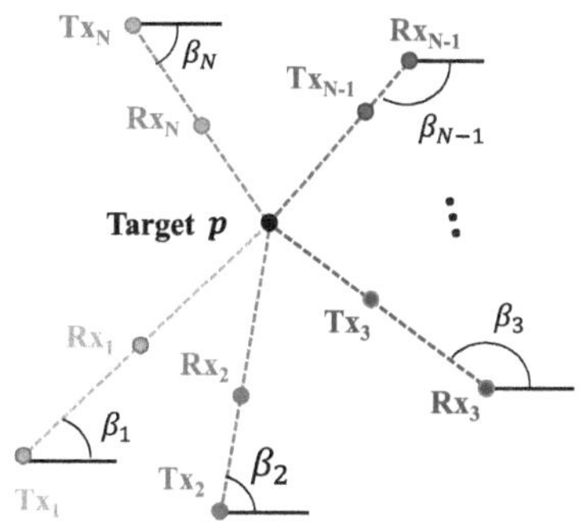

Figure 7.2 Angular separation between different bistatic channels.

Theorem 7.1. *The optimal angular geometries for 2D TOA-based target localization with a multistatic radar system consisting of multiple independent bistatic channels are obtained when the target is collinear with the transmitter and receiver for each bistatic channel with the target being placed at either end, i.e.,* $\theta_{t,i} = \theta_{r,i} = \beta_i$ *for all* $i = 1, 2, \ldots, N$.

The optimization problem now reduces to finding the optimal angular separation between different bistatic channels. The problem of optimal angular separation between different bistatic channels is illustrated in Fig. 7.2. Substituting $\theta_{t,i} = \theta_{r,i} = \beta_i$ into (7.11), we obtain the following corollary as a consequence of Theorem 7.1.

Corollary 7.1. *Maximizing the determinant of the FIM is equivalent to*

$$\max_{\boldsymbol{\beta}} \left\{ \left(\sum_{i=1}^{N} \frac{1}{\sigma_i^2} \right)^2 - \left(\sum_{i=1}^{N} \frac{\sin(2\beta_i)}{\sigma_i^2} \right)^2 - \left(\sum_{i=1}^{N} \frac{\cos(2\beta_i)}{\sigma_i^2} \right)^2 \right\} \tag{7.16}$$

where $\boldsymbol{\beta} = \{\beta_1, \beta_2, \ldots, \beta_N\}$ *and* $\theta_{t,i} = \theta_{r,i} = \beta_i$.

This optimization problem is mathematically identical to that arising in the optimal geometry analysis for AOA localization in [59]. The solutions for (7.16) are outlined as follows.

- $N = 2$:

  The optimal solution is

$$|\beta_1^* - \beta_2^*| = \frac{\pi}{2}. \tag{7.17}$$

  In other words, the optimal angular geometry is obtained when the line of sight (LOS) lines of the two bistatic channels are orthogonal as illustrated in Fig. 7.3.

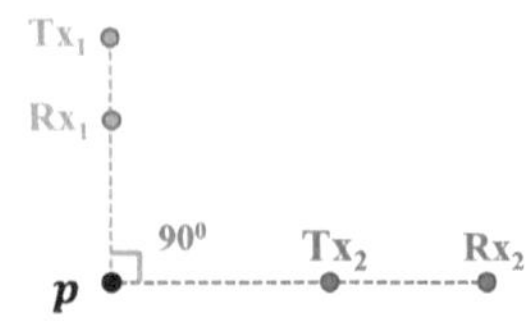

Figure 7.3 An illustration of optimal angular separation for two bistatic channels ($N = 2$).

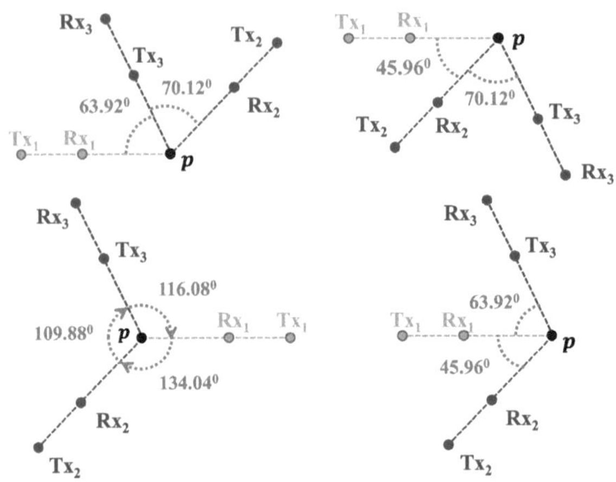

Figure 7.4 An illustration of optimal angular separation for three bistatic channels ($N = 3$) with unequal error variances $\sigma_1 : \sigma_2 : \sigma_3 = 10 : 9 : 8$.

- $N = 3$:
  The optimal solutions are given by

$$\beta_2^* - \beta_1^* = \pm \frac{\tan^{-1}(\sqrt{\zeta}, b^2 - a^2 - 1)}{2}, \tag{7.18a}$$

$$\beta_3^* - \beta_1^* = \mp \frac{\tan^{-1}(\sqrt{\zeta}, a^2 - b^2 - 1)}{2}, \tag{7.18b}$$

where

$$a = \left(\frac{\sigma_1}{\sigma_2}\right)^2, \; b = \left(\frac{\sigma_1}{\sigma_3}\right)^2, \tag{7.19a}$$

$$\zeta = -(a + b + 1)(a - b + 1)(a + b - 1)(a - b - 1). \tag{7.19b}$$

Here, $\tan^{-1}(y, x)$ denotes the four-quadrant inverse tangent of y/x. Four distinct optimal angular geometries are available as illustrated in Fig. 7.4.

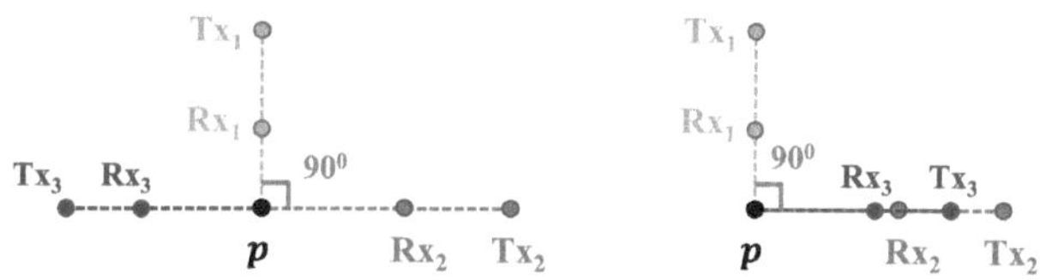

Figure 7.5 An illustration of optimal angular separation for three bistatic channels ($N = 3$) with unequal error variances such that the condition (7.20) satisfies.

However, if $m \in \{1, 2, 3\}$ exists such that

$$\frac{1}{\sigma_m^2} > \sum_{1 \leq i \leq 3, i \neq m} \frac{1}{\sigma_i^2} \tag{7.20}$$

we have $\zeta < 0$. As a result, (7.18) gives complex-valued angles that cannot be used. Instead, the optimal solutions are

$$\beta_k = \pm\frac{\pi}{2} + \beta_m \tag{7.21}$$

with $k \in \{1, 2, 3\} \backslash m$ where $\backslash$ denotes set subtraction. An example for this case is illustrated in Fig. 7.5.

- $N \geq 4$:
  If the condition

$$\frac{2}{\sigma_m^2} > \sum_{i=1}^{N} \frac{1}{\sigma_i^2} \quad \text{with} \quad m = \arg\min_i \sigma_i \tag{7.22}$$

is satisfied, the optimal solutions are

$$\beta_k = \pm\frac{\pi}{2} + \beta_m \tag{7.23}$$

with $k \in \{1, \ldots, N\} \backslash m$. Otherwise, infinitely many solutions exist. The solutions can be determined as in Section 4 of [59]. Alternatively, we can determine the optimal angular configuration by grouping the bistatic channels into smaller clusters (each with two or three channels) and solving the optimal angular separation problem within each cluster [59]. Since the rotation of the optimized clusters does not affect the overall optimality, we can create an infinite number of optimal configurations by rotating one or more optimized clusters [59].

Remark 7.1. For the case of equal error variances $\sigma_1 = \sigma_2 = \cdots = \sigma_N$ ($N \geq 3$), the equal angular separation between different bistatic channels

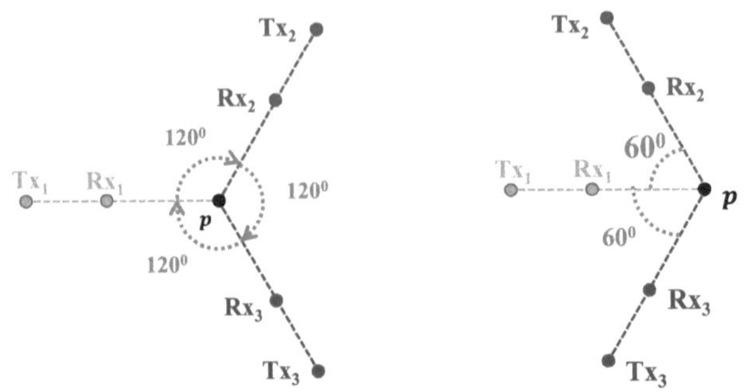

Figure 7.6 An illustration of optimal angular separation for three bistatic channels ($N = 3$) with equal error variance.

is a particular solution among other solutions. An illustration is given in Fig. 7.6.

Remark 7.2. The collinearity of the target with the transmitter and receiver for each bistatic channel does not cause signal blocking since the 2D scenario is an approximation of the actual 3D scenario for low elevation-angle targets.

7.3 Simulation examples

Simulation examples on UAV trajectory optimization are now presented to verify the optimal angular geometries derived in Section 7.2. In the simulations, UAVs act as moving transmitter and receiver platforms to locate a stationary target at $\boldsymbol{p} = [0, 0]^T$ km. Similar to Section 6.4.2, the target position is estimated using the iterative linearized least-squares estimator [74]. To maximize the determinant of the FIM, the gradient-descent trajectory optimization [56] is used to determine the UAV trajectories. Details are given in Section 6.4.2. The TOA error variance at each channel is modeled as range-dependent [75,76], i.e., $E\{e_i^2\} = \varrho_\circ^2 (R_{t,i}^2 R_{r,i}^2)/R_\circ^4$ with $R_{t,i}$ and $R_{r,i}$ being the distances from the target to the transmitter and receiver, respectively. Here, we set the reference range $R_\circ$ to 20 km, the constant $\varrho_\circ$ to 1 μs, and the constant in (28) of [56] to $\Delta = 100$ m. The maximum speed of UAVs is 30 m/s, and the time step is $\Delta_T = 10$ s. Thus, the maximum distance which a UAV can travel in one time step is 300 m. The speed of signal propagation is set to $c = 3 \times 10^8$ m/s. The UAV positions are hard-constrained to maintain a minimum clearance from the target (denoted as $l_{t,i}$ and $l_{r,i}$ for the transmitter and receiver of the ith bistatic channel).

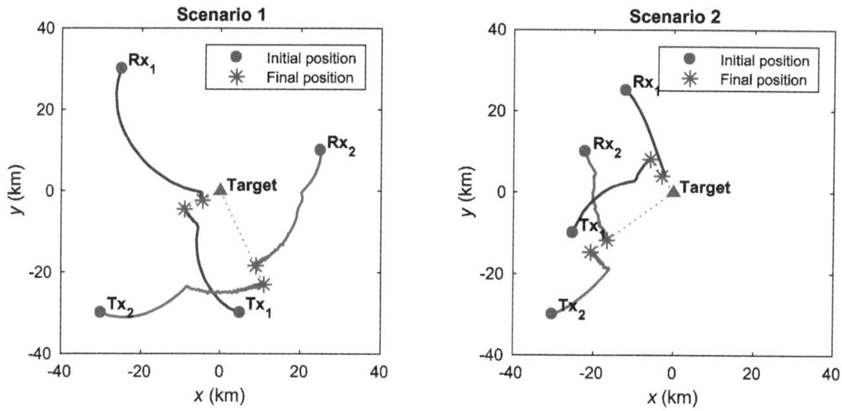

Figure 7.7 Trajectory optimization for two bistatic channels ($N = 2$) with $l_{t,1} = 10$ km, $l_{t,2} = 25$ km, $l_{r,1} = 5$ km, and $l_{r,2} = 20$ km.

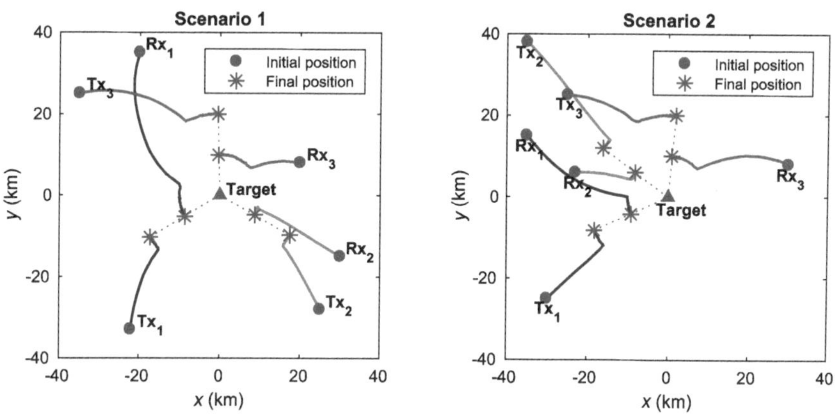

Figure 7.8 Trajectory optimization for three bistatic channels ($N = 3$) with $l_{t,1} = l_{t,2} = l_{t,3} = 20$ km and $l_{r,1} = l_{r,2} = l_{r,3} = 10$ km.

Figs. 7.7–7.10 show the simulated optimal UAV trajectories for two and three bistatic channels with different initial UAV positions. We observe that the UAVs move towards positions that form an optimal angular separation while at the same time approaching the target (because the TOA error variance is reduced when the UAVs get closer to the target). After reaching the minimum distance constraints, the UAVs move around these circular constraints until an optimal angular separation is achieved. Note that, if multiple optimal configurations are available, the final angular configuration formed by the UAVs will depend on the initial UAV positions. In the final

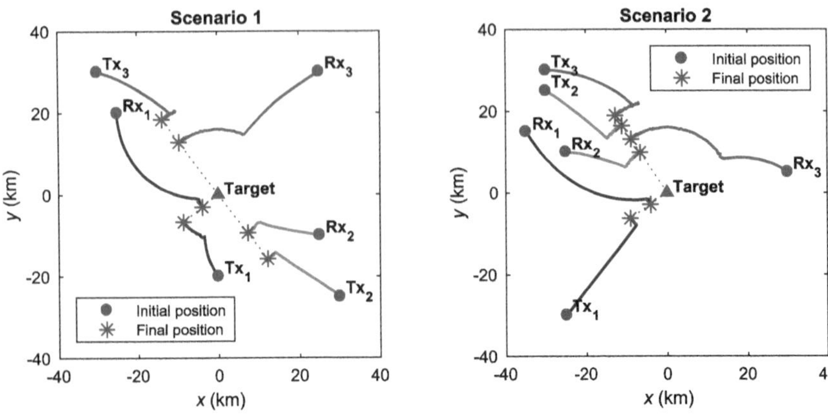

Figure 7.9 Trajectory optimization for three bistatic channels ($N = 3$) with $l_{t,1} = 11$ km, $l_{t,2} = 20$ km, $l_{t,3} = 23$ km, $l_{r,1} = 5$ km, $l_{r,2} = 12$ km, and $l_{r,3} = 16$ km.

UAV formation, the target is collinear with the transmitter and receiver of each bistatic channel (the target is located at either end of the line). This observation confirms the result of Theorem 7.1.

In agreement with the analytical result in Fig. 7.3, we observe that the final angular separation between two bistatic channels is approximately 90° as shown in Fig. 7.7. The simulation results in Fig. 7.8 match the derived angular configurations given Fig. 7.6. Note that, in this scenario, the TOA error variances at different channels are equal when all UAVs reach their hard constraints. On the other hand, when the UAV hard constraints in Fig. 7.9 satisfy the condition in (7.20), the final angular separation of the UAVs matches the optimal angular geometry given in Fig. 7.5. In Fig. 7.10, after all the UAVs arrive at their hard constraints, the TOA error variance ratios are $\sigma_1 : \sigma_2 : \sigma_3 = 10 : 9 : 8$, which is identical to that in Fig. 7.4. We observe that the simulated results in Fig. 7.10 match the analytical findings in Fig. 7.4.

7.4 Summary

This chapter has presented an optimal geometry analysis for the 2D TOA localization problem using a multistatic system with independent bistatic channels. The optimal geometries were analytically derived based on maximizing the determinant of the FIM, which effectively minimizes the area of the estimation confidence region. The optimal angular geometries are obtained when the target is collinear with the transmitter and receiver for

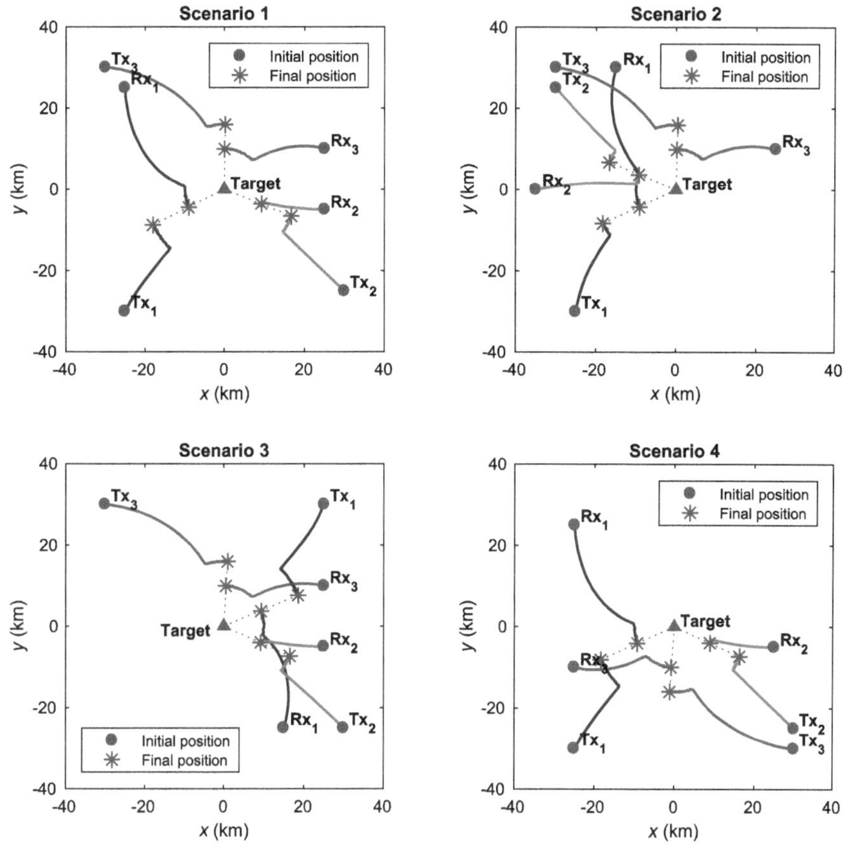

Figure 7.10 Trajectory optimization for three bistatic channels ($N = 3$) with $l_{t,1} = 20$ km, $l_{t,2} = 18$ km, $l_{t,3} = 16$ km, and $l_{r,1} = l_{r,2} = l_{r,3} = 10$ km.

each bistatic channel with the target being located at either end of the line. The problem of finding an optimal geometry then becomes one of optimizing the angular separation between different bistatic channels which can be readily solved building on the results of optimal angular separation for AOA localization. The optimal angular separation between two bistatic channels is obtained when their LOS lines are perpendicular. A finite number of optimal angular separation solutions are available for three bistatic channels, which are governed by the noise variances. For a multistatic system with four or more bistatic channels, infinitely many optimal angular separation configurations may exist and can be determined by forming several clusters, each with two or three channels, and optimizing the angular separation within each cluster. Simulation examples on UAV trajectory optimization were presented to verify the derived optimal geometries, where good agreement between the analytical and numerical results was observed.

PART 3

Pseudolinear tracking algorithms

Background and purpose

The main challenge for target tracking and localization with multistatic radar systems is the nonlinear relationship between the noisy radar measurements (e.g., AOA, TDOA and FDOA) and the target kinematic parameters (position and velocity). Nonlinear least-squares approaches like the maximum likelihood estimator are not only computationally expensive, but also prone to convergence to local minima or divergence due to iterative numerical computation and the threshold effect. To overcome such shortcomings, the nonlinear measurement equations can be algebraically rearranged into a set of equations that are linear in the unknowns, thus facilitating the use of linear least-squares in closed-form. This part of the book is devoted to the design and analysis of inherently-stable and low-complexity closed-form algorithms for passive multistatic tracking and localization. Specifically, Chapter 8 presents batch pseudolinear estimators for multistatic target motion analysis of a constant-velocity target utilizing AOA, TDOA and FDOA measurements, while Chapter 9 covers a range of algebraic closed-form estimators for multistatic localization with TDOA measurements.

CHAPTER 8

Batch track estimators for multistatic target motion analysis

Contents

8.1 Introduction

Target motion analysis (TMA) has been an active research area for many years owing to its various civilian and military applications in acoustic source localization, wireless sensor networks, and radar and sonar systems, to name a few. The objective of TMA is to estimate the kinematic parameters (i.e., position, velocity and possibly acceleration) of a moving target from noisy measurements collected by a moving sensor, or several spatially distributed stationary or moving sensors. Different to target tracking where recursive tracking algorithms, such as the Kalman filter and its variants, are often employed, TMA usually considers a deterministic model for target dynamics (e.g., a constant-velocity or -acceleration model), thus allowing the target kinematic parameters to be estimated in a batch-processing manner. In this chapter, we focus our attention on the problem of TMA for a constant-velocity target by a passive multistatic radar system using AOA, TDOA and FDOA measurements.

Estimating the target kinematic parameters from AOA, TDOA and FDOA measurements is not trivial because of the nonlinear nature of these

Signal Processing for Multistatic Radar Systems
https://doi.org/10.1016/B978-0-12-815314-7.00018-4

measurements with respect to the target kinematic parameters. In spite of enjoying the desirable properties of being asymptotically unbiased and efficient, the MLE does not have a closed-form solution and thus requires the use of iterative numerical search algorithms. For this reason, the MLE is not only computationally expensive but also vulnerable to instability, which also follows from the non-convexity of the maximum-likelihood cost function. An attractive approach to overcome these shortcomings of the MLE is to rearrange the nonlinear equations of AOA, TDOA and FDOA into a linear form so that closed-form linear least-squares can be applied. In the context of AOA-only TMA with a moving observer, this technique is commonly referred to as the PLE [77,78]. Being closed-form makes the PLE not only computationally simple, but also avoids the instability problems associated with the iterative solutions of the MLE. A drawback of the PLE is that it produces biased estimates due to the correlation between the measurement matrix and pseudolinear noise vector in the pseudolinear equation [77–79]. To overcome such a bias problem, the use of bias compensation and instrument variable (IV) estimation was proposed in [80,81] for AOA-only TMA. The method of IV estimation was also used in [82] for the problem of single-platform Doppler-bearing TMA but in an iterative manner, and in [83–85] for recursive target tracking with Kalman filtering. In [86], a two-stage algebraic solution was developed for localizing a moving target using TDOA and FDOA measurements. However, this work estimates the target position and velocity from TDOA and FDOA measurements for each single time instant separately without taking into account information on the target dynamics.

This chapter presents closed-form pseudolinear estimators for passive multistatic TMA using AOA, TDOA and FDOA measurements. To develop the PLE, the TDOA and FDOA equations are linearized with the help of AOA information. This linearization approach avoids the appearance of undesirable nuisance parameters in the unknown vector and thus excludes the need for extra effort to tackle the dependence of the nuisance parameters on the target motion parameters. To alleviate the bias problems associated with the PLE, an asymptotic bias property of the PLE is analyzed and a bias-compensated variant of the PLE, called BCPLE, is presented based on instantaneous bias estimation. The idea of IV estimation is then exploited to develop the WIVE which enjoys the asymptotically unbiased property. The WIVE is also shown to be efficient in the asymptotic sense, i.e., achieving the CRLB, under the small noise assumption.

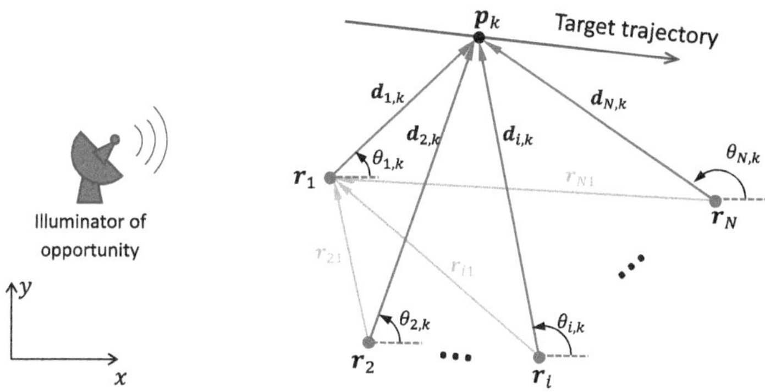

Figure 8.1 TMA geometry with a passive multistatic radar system.

8.2 Problem formulation

Fig. 8.1 depicts the TMA problem using a passive multistatic radar system in the 2D plane. The multistatic radar system consists of N receivers located at $\boldsymbol{r}_i = [r_{x,i}, r_{y,i}]^T$ $(i = 1, \ldots, N)$. The receivers exploit a nearby radar or television/radio transmitter as the illuminator of opportunity. The receivers passively collect AOA, TDOA and FDOA measurements which are used to track a moving target. Here, $\boldsymbol{p}_k = [p_{x,k}, p_{y,k}]^T$ and $\boldsymbol{v}_k = [v_{x,k}, v_{y,k}]^T$ denote the unknown position and velocity of the target, respectively, at time instant $k \in \{0, 1, \ldots, M-1\}$. The objective of the TMA problem is to estimate the target position $\boldsymbol{p}_k$ and velocity $\boldsymbol{v}_k$ using noisy AOA, TDOA and FDOA measurements collected at the receivers during the entire observation period, $k = 0, \ldots, M-1$, in a batch-processing manner. For the TMA problem under consideration, a minimum of two receivers are sufficient for ensuring the uniqueness of the solution.

The true AOA at receiver i at time instant k as a function of the target position $\boldsymbol{p}_k$ and receiver position $\boldsymbol{r}_i$ is given by

$$\theta_{i,k} = \tan^{-1} \frac{\Delta y_{i,k}}{\Delta x_{i,k}} \tag{8.1}$$

where $\Delta x_{i,k} = p_{x,k} - r_{x,i}$, $\Delta y_{i,k} = p_{y,k} - r_{y,i}$. The true TDOA between the signals arriving at receivers i and j at time instant k is related to $\boldsymbol{p}_k$ and $\boldsymbol{r}_i$ via

$$\tau_{ji,k} = \tau_{j,k} - \tau_{i,k} \tag{8.2}$$

where $\tau_{i,k} = \|\boldsymbol{d}_{i,k}\|/c$ denotes the duration of the signal traveling from the target to receiver i. Here, we have defined $\boldsymbol{d}_{i,k} = \boldsymbol{p}_k - \boldsymbol{r}_i$. Multiplying (8.2) with c, we obtain the difference between the distances from the target to receivers i and j:

$$\gamma_{ji,k} = \|\boldsymbol{d}_{j,k}\| - \|\boldsymbol{d}_{i,k}\|. \tag{8.3}$$

The true FDOA between the signals arriving at receivers i and j at time instant k is given by

$$f_{ji,k} = \frac{f_\circ}{c}\left(\dot{d}_{j,k} - \dot{d}_{i,k}\right) = \frac{f_\circ}{c}\left(\frac{\boldsymbol{d}_{j,k}^T \boldsymbol{v}_k}{\|\boldsymbol{d}_{j,k}\|} - \frac{\boldsymbol{d}_{i,k}^T \boldsymbol{v}_k}{\|\boldsymbol{d}_{i,k}\|}\right) \tag{8.4}$$

where $f_\circ$ is the transmission frequency and $d_{i,k} = \|\boldsymbol{d}_{i,k}\|$. Since the radar system under consideration is passive, the transmission frequency is unknown by the receivers. However, this frequency can be readily estimated using the LOS signal from the transmitter arriving at one of the receivers. In this chapter, the transmission frequency is assumed to be estimated and known *a priori*. By normalizing (8.4) with $f_\circ/c$, the FDOA becomes

$$\zeta_{ji,k} = \frac{\boldsymbol{d}_{j,k}^T \boldsymbol{v}_k}{\|\boldsymbol{d}_{j,k}\|} - \frac{\boldsymbol{d}_{i,k}^T \boldsymbol{v}_k}{\|\boldsymbol{d}_{i,k}\|}. \tag{8.5}$$

It is common practice to utilize one of the receivers as a reference receiver for TDOA and FDOA measurements. Receiver 1 is chosen here as the reference receiver. Thus, there are a total of $N - 1$ TDOA measurements $\gamma_{i1,k}$ and $N - 1$ FDOA measurements $\zeta_{i1,k}$ for $i = 2, \ldots, N$ at each instant k.

Throughout this chapter, we make the following assumptions:

- The target follows the constant-velocity kinematic model during the observation interval $k \in \{0, \ldots, M - 1\}$. The extension to the constant-acceleration model is straightforward. In many applications, these models are sufficiently good to represent the actual target dynamics for a short observation interval [77,78,80,81,87,88]. If the observation interval is long, it can be split into several smaller intervals, thus allowing the batch TMA algorithms to be applied to each sub-interval.
- The radar measurements are obtained at regular time instants $t_k = kT/(M-1)$, $k = 0, \ldots, M-1$, where T is the total observation duration. Following the constant-velocity kinematic model, the target position $\boldsymbol{p}_k$ and velocity $\boldsymbol{v}_k$ at time instant t_k are

$$\boldsymbol{p}_k = \boldsymbol{p}_0 + t_k \boldsymbol{v}_0 = \boldsymbol{M}_k \boldsymbol{\xi}, \tag{8.6a}$$

$$\boldsymbol{v}_k = \boldsymbol{v}_0 = \boldsymbol{L}\boldsymbol{\xi}, \tag{8.6b}$$

where

$$\boldsymbol{M}_k = \begin{bmatrix} 1 & 0 & t_k & 0 \\ 0 & 1 & 0 & t_k \end{bmatrix}, \quad \boldsymbol{L} = \begin{bmatrix} 0 & 0 & 1 & 0 \\ 0 & 0 & 0 & 1 \end{bmatrix}, \quad \boldsymbol{\xi} = \begin{bmatrix} \boldsymbol{p}_0 \\ \boldsymbol{v}_0 \end{bmatrix}. \tag{8.7}$$

Here, $\boldsymbol{\xi}$ is a 4×1 vector of *target motion parameters* to be estimated. Once $\boldsymbol{\xi}$ is estimated, we can determine the target position $\boldsymbol{p}_k$ and velocity $\boldsymbol{v}_k$ at any time instant k using simple substitutions into (8.6).

- The radar measurements are corrupted by additive noise:

$$\tilde{\theta}_{i,k} = \theta_{i,k} + n_{i,k}, \quad i = 1, \ldots, N, \tag{8.8a}$$
$$\tilde{\gamma}_{i1,k} = \gamma_{i1,k} + \varepsilon_{i1,k}, \quad i = 2, \ldots, N, \tag{8.8b}$$
$$\tilde{\zeta}_{i1,k} = \zeta_{i1,k} + \varrho_{i1,k}, \quad i = 2, \ldots, N, \tag{8.8c}$$

where $n_{i,k}$, $\varepsilon_{i1,k}$ and $\varrho_{i1,k}$ are noise terms which are modeled as zero-mean Gaussian random variables. Stacking all radar measurements at time instant k into a vector yields

$$\tilde{\boldsymbol{\psi}}_k = \boldsymbol{\psi}_k + \boldsymbol{\eta}_k \tag{8.9}$$

where

$$\tilde{\boldsymbol{\psi}}_k = [\tilde{\theta}_{1,k}, \ldots, \tilde{\theta}_{N,k}, \tilde{\gamma}_{21,k}, \ldots, \tilde{\gamma}_{N1,k}, \tilde{\zeta}_{21,k}, \ldots, \tilde{\zeta}_{N1,k}]^T, \tag{8.10a}$$
$$\boldsymbol{\psi}_k = [\theta_{1,k}, \ldots, \theta_{N,k}, \gamma_{21,k}, \ldots, \gamma_{N1,k}, \zeta_{21,k}, \ldots, \zeta_{N1,k}]^T, \tag{8.10b}$$
$$\boldsymbol{\eta}_k = [n_{1,k}, \ldots, n_{N,k}, \varepsilon_{21,k}, \ldots, \varepsilon_{N1,k}, \varrho_{21,k}, \ldots, \varrho_{N1,k}]^T. \tag{8.10c}$$

The noise vector $\boldsymbol{\eta}_k$ has a covariance matrix in the block-diagonal form given by

$$\boldsymbol{K}_k = \mathbb{E}\left\{\boldsymbol{\eta}_k \boldsymbol{\eta}_k^T\right\} = \operatorname{diag}\left(\boldsymbol{K}_k^{\mathrm{AOA}}, \boldsymbol{K}_k^{\mathrm{TDOA/FDOA}}\right) \tag{8.11}$$

where

$$\boldsymbol{K}_k^{\mathrm{AOA}} = \operatorname{diag}\left(\sigma_{\theta,1,k}^2, \ldots, \sigma_{\theta,N,k}^2\right) \tag{8.12}$$

and

$$\boldsymbol{K}_k^{\mathrm{TDOA/FDOA}} = \begin{bmatrix} \boldsymbol{K}_{\varepsilon\varepsilon,k} & \boldsymbol{K}_{\varepsilon\varrho,k} \\ \boldsymbol{K}_{\varepsilon\varrho,k}^T & \boldsymbol{K}_{\varrho\varrho,k} \end{bmatrix}. \tag{8.13}$$

The AOA error is assumed to be uncorrelated with the TDOA/FDOA errors [82,89–92]. This assumption perfectly holds if the AOA estimation is performed separately from the TDOA/FDOA estimation. If the

AOA is estimated jointly with TDOA/FDOA, this is only an approximation. Here, $\boldsymbol{K}_{\varepsilon\varrho,k}$ accounts for the cross-correlation between the TDOA and FDOA errors. Note that $\sigma_{\theta,i,k}$, $\boldsymbol{K}_{\varepsilon\varepsilon,k}$, $\boldsymbol{K}_{\varrho\varrho,k}$ and $\boldsymbol{K}_{\varepsilon\varrho,k}$ can vary with time k and are assumed to be known *a priori*.

- The radar measurement errors are statistically independent over time k, i.e., $E\{\boldsymbol{\eta}_k\boldsymbol{\eta}_h^T\}=\mathbf{0}$ for $k\neq h$. Stacking (8.9) for $k=0,\ldots,M-1$ gives

$$\tilde{\boldsymbol{\psi}}=\boldsymbol{\psi}+\boldsymbol{\eta} \tag{8.14}$$

where

$$\tilde{\boldsymbol{\psi}}=\left[\tilde{\boldsymbol{\psi}}_0^T,\tilde{\boldsymbol{\psi}}_1^T,\ldots,\tilde{\boldsymbol{\psi}}_{M-1}^T\right]^T, \tag{8.15a}$$

$$\boldsymbol{\psi}=\left[\boldsymbol{\psi}_0^T,\boldsymbol{\psi}_1^T,\ldots,\boldsymbol{\psi}_{M-1}^T\right]^T, \tag{8.15b}$$

$$\boldsymbol{\eta}=\left[\boldsymbol{\eta}_0^T,\boldsymbol{\eta}_1^T,\ldots,\boldsymbol{\eta}_{M-1}^T\right]^T, \tag{8.15c}$$

and the covariance matrix of $\boldsymbol{\eta}$ is given by

$$\boldsymbol{K}=\mathbb{E}\left\{\boldsymbol{\eta}\boldsymbol{\eta}^T\right\}=\operatorname{diag}\left(\boldsymbol{K}_0,\boldsymbol{K}_1,\ldots,\boldsymbol{K}_{M-1}\right). \tag{8.16}$$

8.3 Maximum likelihood estimator and Cramér–Rao lower bound

The likelihood function for the measurement vector $\tilde{\boldsymbol{\psi}}$ under the Gaussian noise assumption is a multivariate Gaussian probability density function given by

$$\mathcal{L}(\tilde{\boldsymbol{\psi}}|\boldsymbol{\xi})=\frac{1}{(2\pi)^{\frac{M(3N-2)}{2}}|\boldsymbol{K}|^{\frac{1}{2}}}\exp\left\{-\frac{1}{2}(\tilde{\boldsymbol{\psi}}-\boldsymbol{\psi}(\boldsymbol{\xi}))^T\boldsymbol{K}^{-1}(\tilde{\boldsymbol{\psi}}-\boldsymbol{\psi}(\boldsymbol{\xi}))\right\}. \tag{8.17}$$

Note that the true AOA $\theta_{i,k}$, TDOA $\gamma_{i1,k}$ and FDOA $\zeta_{i1,k}$ are dependent on the target motion parameter vector $\boldsymbol{\xi}$ through

$$\theta_{i,k}(\boldsymbol{\xi})=\tan^{-1}\frac{\Delta y_{i,k}(\boldsymbol{\xi})}{\Delta x_{i,k}(\boldsymbol{\xi})},\quad i=1,\ldots,N, \tag{8.18a}$$

$$\gamma_{i1,k}(\boldsymbol{\xi})=\|\boldsymbol{d}_{i,k}(\boldsymbol{\xi})\|-\|\boldsymbol{d}_{1,k}(\boldsymbol{\xi})\|,\quad i=2,\ldots,N, \tag{8.18b}$$

$$\zeta_{i1,k}(\boldsymbol{\xi})=\frac{\boldsymbol{d}_{i,k}^T(\boldsymbol{\xi})\boldsymbol{v}_k(\boldsymbol{\xi})}{\|\boldsymbol{d}_{i,k}(\boldsymbol{\xi})\|}-\frac{\boldsymbol{d}_{1,k}^T(\boldsymbol{\xi})\boldsymbol{v}_k(\boldsymbol{\xi})}{\|\boldsymbol{d}_{1,k}(\boldsymbol{\xi})\|},\quad i=2,\ldots,N, \tag{8.18c}$$

where

$$\boldsymbol{d}_{i,k}(\boldsymbol{\xi}) = \begin{bmatrix} \Delta x_{i,k}(\boldsymbol{\xi}) \\ \Delta y_{i,k}(\boldsymbol{\xi}) \end{bmatrix} = \boldsymbol{M}_k \boldsymbol{\xi} - \boldsymbol{r}_i \tag{8.19}$$

for $k = 0, \ldots, M-1$. As a result, $\boldsymbol{\psi}(\boldsymbol{\xi})$ is explicitly written as a function of $\boldsymbol{\xi}$.

The maximum-likelihood estimate of $\boldsymbol{\xi}$ (denoted as $\hat{\boldsymbol{\xi}}_{\text{MLE}}$) is obtained by maximizing the log-likelihood function $\ln \mathcal{L}(\tilde{\boldsymbol{\psi}}|\boldsymbol{\xi})$ over $\boldsymbol{\xi}$, which can be equivalently written as

$$\hat{\boldsymbol{\xi}}_{\text{MLE}} = \arg\min_{\boldsymbol{\xi}\in\mathbb{R}^4} J_{\text{ML}}(\boldsymbol{\xi}) \tag{8.20}$$

where $J_{\text{ML}}(\boldsymbol{\xi})$ is the maximum-likelihood cost function

$$J_{\text{ML}}(\boldsymbol{\xi}) = \frac{1}{2}\boldsymbol{\vartheta}^T(\boldsymbol{\xi})\boldsymbol{K}^{-1}\boldsymbol{\vartheta}(\boldsymbol{\xi}), \quad \boldsymbol{\vartheta}(\boldsymbol{\xi}) = \tilde{\boldsymbol{\psi}} - \boldsymbol{\psi}(\boldsymbol{\xi}). \tag{8.21}$$

The minimization problem in (8.20) is in fact a nonlinear least-squares estimation problem which does not admit a closed-form solution. A numerical solution of (8.20) can be obtained via various gradient or downhill simplex-based iterative search algorithms such as the Gauss–Newton algorithm [93], the steepest-descent algorithm [94], and the Nelder–Mead simplex algorithm [95]. In particular, the Gauss–Newton iterations take the following form:

$$\hat{\boldsymbol{\xi}}(j+1) = \hat{\boldsymbol{\xi}}(j) + (\boldsymbol{J}^T(j)\boldsymbol{K}^{-1}\boldsymbol{J}(j))^{-1}\boldsymbol{J}^T(j)\boldsymbol{K}^{-1}\boldsymbol{\vartheta}(\hat{\boldsymbol{\xi}}(j)), \quad j = 0, 1, \ldots \tag{8.22}$$

where $\boldsymbol{J}(j)$ is the Jacobian matrix of $\boldsymbol{\psi}(\boldsymbol{\xi})$ with respect to $\boldsymbol{\xi}$ evaluated at $\boldsymbol{\xi} = \hat{\boldsymbol{\xi}}(j)$:

$$\boldsymbol{J}(j) = [\boldsymbol{J}_0^T(j), \boldsymbol{J}_1^T(j), \ldots, \boldsymbol{J}_{M-1}^T(j)]^T \tag{8.23}$$

with

$$\boldsymbol{J}_k(j) = [\boldsymbol{J}_{\theta_{1,k}}^T(j), \ldots, \boldsymbol{J}_{\theta_{N,k}}^T(j), \boldsymbol{J}_{\gamma_{21,k}}^T(j), \ldots, \boldsymbol{J}_{\gamma_{N1,k}}^T(j), \boldsymbol{J}_{\zeta_{21,k}}^T(j), \ldots, \boldsymbol{J}_{\zeta_{N1,k}}^T(j)]^T. \tag{8.24}$$

The expressions for $\boldsymbol{J}_{\theta_{i,k}}(j)$, $\boldsymbol{J}_{\gamma_{i1,k}}(j)$ and $\boldsymbol{J}_{\zeta_{i1,k}}(j)$ are

$$\boldsymbol{J}_{\theta_{i,k}}(j) = \frac{\left[-\sin\theta_{i,k}(\hat{\boldsymbol{\xi}}(j)), \cos\theta_{i,k}(\hat{\boldsymbol{\xi}}(j))\right]\boldsymbol{M}_k}{\|\boldsymbol{d}_{i,k}(\hat{\boldsymbol{\xi}}(j))\|}, \tag{8.25a}$$

$$J_{\gamma_{i1,k}}(j) = \begin{bmatrix} \cos\theta_{i,k}(\hat{\boldsymbol{\xi}}(j)) - \cos\theta_{1,k}(\hat{\boldsymbol{\xi}}(j)) \\ \sin\theta_{i,k}(\hat{\boldsymbol{\xi}}(j)) - \sin\theta_{1,k}(\hat{\boldsymbol{\xi}}(j)) \end{bmatrix}^T \boldsymbol{M}_k, \tag{8.25b}$$

$$\begin{aligned} J_{\zeta_{i1,k}}(j) = {} & \begin{bmatrix} \cos\theta_{i,k}(\hat{\boldsymbol{\xi}}(j)) - \cos\theta_{1,k}(\hat{\boldsymbol{\xi}}(j)) \\ \sin\theta_{i,k}(\hat{\boldsymbol{\xi}}(j)) - \sin\theta_{1,k}(\hat{\boldsymbol{\xi}}(j)) \end{bmatrix}^T \boldsymbol{L} \\ & + \left(\frac{\hat{\xi}_4(j)\cos\theta_{i,k}(\hat{\boldsymbol{\xi}}(j)) - \hat{\xi}_3(j)\sin\theta_{i,k}(\hat{\boldsymbol{\xi}}(j))}{\|\boldsymbol{d}_{i,k}(\hat{\boldsymbol{\xi}}(j))\|} \right. \\ & \qquad \left. \times \left[-\sin\theta_{i,k}(\hat{\boldsymbol{\xi}}(j)), \cos\theta_{i,k}(\hat{\boldsymbol{\xi}}(j)) \right] \boldsymbol{M}_k \right) \\ & - \left(\frac{\hat{\xi}_4(j)\cos\theta_{1,k}(\hat{\boldsymbol{\xi}}(j)) - \hat{\xi}_3(j)\sin\theta_{1,k}(\hat{\boldsymbol{\xi}}(j))}{\|\boldsymbol{d}_{1,k}(\hat{\boldsymbol{\xi}}(j))\|} \right. \\ & \qquad \left. \times \left[-\sin\theta_{1,k}(\hat{\boldsymbol{\xi}}(j)), \cos\theta_{1,k}(\hat{\boldsymbol{\xi}}(j)) \right] \boldsymbol{M}_k \right). \end{aligned} \tag{8.25c}$$

Note that $\hat{\xi}_3(j)$ and $\hat{\xi}_4(j)$ denote the third and fourth elements of $\hat{\boldsymbol{\xi}}(j)$ which correspond to the velocity components.

Although the MLE enjoys certain desirable estimation properties such as asymptotic unbiasedness and efficiency, the iterative MLE solution is not only computationally demanding but also susceptible to divergence due to the nonconvex nature of the maximum-likelihood cost function. At high noise levels, it is also subject to the threshold effect, which manifests itself as a rapid deterioration of the MSE performance when the noise increases [96]. For this reason, the MLE is often used as a performance benchmark for comparison purposes.

The CRLB for the TMA problem is

$$\boldsymbol{C}_{\xi} = (\boldsymbol{J}_{\circ}^T \boldsymbol{K}^{-1} \boldsymbol{J}_{\circ})^{-1} \tag{8.26}$$

where $\boldsymbol{J}_{\circ}$ is the Jacobian matrix calculated at the true value of $\boldsymbol{\xi}$.

8.4 Pseudolinear estimator

The main motivation for pseudolinear estimation is to algebraically rearrange the nonlinear AOA, TDOA and FDOA measurement equations so as to make them linear in the unknown target motion parameter vector $\boldsymbol{\xi}$, thus enabling the use of closed-form linear least-squares estimation. In this section, we derive the pseudolinear estimator by showing how the measurement equations are linearized in the unknowns and a linear matrix equation is constructed.

8.4.1 Pseudolinear equations

AOA

Using (8.1) and (8.8a), we obtain

$$\frac{\sin(\tilde{\theta}_{i,k} - n_{i,k})}{\cos(\tilde{\theta}_{i,k} - n_{i,k})} = \frac{\Delta y_{i,k}}{\Delta x_{i,k}}. \tag{8.27}$$

After some algebraic manipulations, (8.27) can be rewritten as the pseudo-linear equation for the AOA measurement $\tilde{\theta}_{i,k}$ [87]:

$$\boldsymbol{F}_{\theta_{i,k}} \boldsymbol{\xi} = b_{\theta_{i,k}} + e_{\theta_{i,k}} \tag{8.28}$$

where

$$\boldsymbol{F}_{\theta_{i,k}} = \left[\sin\tilde{\theta}_{i,k}, \quad -\cos\tilde{\theta}_{i,k}\right] \boldsymbol{M}_k, \tag{8.29a}$$

$$b_{\theta_{i,k}} = \left[\sin\tilde{\theta}_{i,k}, \quad -\cos\tilde{\theta}_{i,k}\right] \boldsymbol{r}_i, \tag{8.29b}$$

$$e_{\theta_{i,k}} = \|\boldsymbol{d}_{i,k}\| \sin n_{i,k} \approx \|\boldsymbol{d}_{i,k}\| n_{i,k}. \tag{8.29c}$$

Eq. (8.28) can also be derived by exploiting an orthogonal vector sum relationship between $\boldsymbol{p}_k$ and $\boldsymbol{r}_i$ [97,98], or using a small-noise approximation of the MLE cost function [99].

Note that, for sufficiently small noise, the second- and higher-order terms of the Taylor series expansion of $\sin n_{i,k}$ can be ignored, thus leading to the approximation of $\sin n_{i,k} \approx n_{i,k}$ as used in (8.29c).

TDOA

The nonlinear TDOA measurement equation is linearized with the support of the AOA measurement to avoid the creation of a dependent nuisance parameter. Using the law of cosines applied to the triangle formed by the corner points $\boldsymbol{p}_k$, $\boldsymbol{r}_1$ and $\boldsymbol{r}_i$, we have

$$\|\boldsymbol{d}_{1,k}\|^2 = \|\boldsymbol{d}_{i,k}\|^2 + \|\boldsymbol{r}_{i1}\|^2 - 2\boldsymbol{r}_{i1}^T \boldsymbol{d}_{i,k} \tag{8.30}$$

where $\boldsymbol{r}_{i1} = \boldsymbol{r}_1 - \boldsymbol{r}_i$ (i.e., the vector pointing from receiver i to receiver 1). From (8.3), we obtain

$$\|\boldsymbol{d}_{1,k}\|^2 = \|\boldsymbol{d}_{i,k}\|^2 - 2\gamma_{i1,k}\|\boldsymbol{d}_{i,k}\| + \gamma_{i1,k}^2. \tag{8.31}$$

Recognizing (8.30) and (8.31) are identical, we obtain

$$\gamma_{i1,k}\|\boldsymbol{d}_{i,k}\| = -\frac{1}{2}\|\boldsymbol{r}_{i1}\|^2 + \frac{1}{2}\gamma_{i1,k}^2 + \boldsymbol{r}_{i1}^T\boldsymbol{d}_{i,k}. \tag{8.32}$$

From the inner product of $\boldsymbol{r}_{i1,k}$ and $\boldsymbol{d}_{i,k}$

$$\boldsymbol{r}_{i1}^T\boldsymbol{d}_{i,k} = \|\boldsymbol{r}_{i1,k}\|\,\|\boldsymbol{d}_{i,k}\|\cos(\theta_{i,k} - \theta_{i1}) \tag{8.33}$$

we have

$$\|\boldsymbol{d}_{i,k}\| = \frac{\boldsymbol{r}_{i1}^T\boldsymbol{d}_{i,k}}{\|\boldsymbol{r}_{i1,k}\|\cos(\theta_{i,k} - \theta_{i1})} \tag{8.34}$$

where θ_{i1} is the bearing angle of $\boldsymbol{r}_{i1}$, defined by

$$\theta_{i1} = \tan^{-1}\left(\frac{r_{y,1} - r_{y,i}}{r_{x,1} - r_{x,i}}\right). \tag{8.35}$$

Substituting (8.34) into (8.32) gives

$$\left(\frac{\gamma_{i1,k}}{\|\boldsymbol{r}_{i1}\|} - \cos(\theta_{i,k} - \theta_{i1})\right)\boldsymbol{r}_{i1}^T\boldsymbol{d}_{i,k} = \left(-\frac{1}{2}\|\boldsymbol{r}_{i1}\|^2 + \frac{1}{2}\gamma_{i1,k}^2\right)\cos(\theta_{i,k} - \theta_{i1}). \tag{8.36}$$

Plugging $\boldsymbol{d}_{i,k} = \boldsymbol{M}_k\boldsymbol{\xi} - \boldsymbol{r}_i$ into (8.36) and replacing the true TDOA $\gamma_{i1,k}$ and AOA $\theta_{i,k}$ with their corresponding noisy measurements $\tilde{\gamma}_{i1,k}$ and $\tilde{\theta}_{i,k}$, respectively, we obtain the TDOA pseudolinear equation as

$$\boldsymbol{F}_{\gamma_{i1,k}}\boldsymbol{\xi} = b_{\gamma_{i1,k}} + e_{\gamma_{i1,k}} \tag{8.37}$$

where

$$\boldsymbol{F}_{\gamma_{i1,k}} = \left(\frac{\tilde{\gamma}_{i1,k}}{\|\boldsymbol{r}_{i1}\|} - \cos(\tilde{\theta}_{i,k} - \theta_{i1})\right)\frac{\boldsymbol{r}_{i1}^T\boldsymbol{M}_k}{\|\boldsymbol{r}_{i1}\|}, \tag{8.38a}$$

$$b_{\gamma_{i1,k}} = \left(-\frac{1}{2}\|\boldsymbol{r}_{i1}\|^2 + \frac{1}{2}\tilde{\gamma}_{i1,k}^2\right)\frac{\cos(\tilde{\theta}_{i,k} - \theta_{i1})}{\|\boldsymbol{r}_{i1}\|} + \left(\frac{\tilde{\gamma}_{i1,k}}{\|\boldsymbol{r}_{i1}\|} - \cos(\tilde{\theta}_{i,k} - \theta_{i1})\right)\frac{\boldsymbol{r}_{i1}^T\boldsymbol{r}_i}{\|\boldsymbol{r}_{i1}\|}, \tag{8.38b}$$

$$e_{\gamma_{i1,k}} \approx \frac{\|\boldsymbol{d}_{1,k}\|}{\|\boldsymbol{r}_{i1}\|}\cos(\theta_{i,k} - \theta_{i1})\varepsilon_{i1,k} + \frac{\|\boldsymbol{d}_{i,k}\|\left(\|\boldsymbol{d}_{i,k}\| - \|\boldsymbol{d}_{1,k}\|\right)}{\|\boldsymbol{r}_{i1}\|}\sin(\theta_{i,k} - \theta_{i1})n_{i,k}. \tag{8.38c}$$

The pseudolinear noise $e_{\gamma_{i1,k}}$ in (8.38c) is obtained by ignoring the second and higher-order noise terms under the small noise approximation. An exact expression for $e_{\gamma_{i1,k}}$ is given in Section 8.11 (Appendix B).

FDOA

Similar to TDOA, the pseudolinear equation for FDOA can also be formulated with the help of AOA measurements. From (8.5), we have

$$\zeta_{i1,k} = \left(\frac{\boldsymbol{d}_{i,k}^T}{\|\boldsymbol{d}_{i,k}\|} - \frac{\boldsymbol{d}_{1,k}^T}{\|\boldsymbol{d}_{1,k}\|} \right) \boldsymbol{v}_k = \left[\cos\theta_{i,k} - \cos\theta_{1,k},\ \sin\theta_{i,k} - \sin\theta_{1,k}\right] \boldsymbol{v}_k. \tag{8.39}$$

Noting $\boldsymbol{v}_k = \boldsymbol{L}\boldsymbol{\xi}$ and replacing the noise-free terms $\zeta_{i1,k}$, $\theta_{i,k}$ and $\theta_{1,k}$ with their respective noisy measurements $\tilde{\zeta}_{i1,k}$, $\tilde{\theta}_{i,k}$ and $\tilde{\theta}_{1,k}$, (8.39) can be re-expressed as

$$\boldsymbol{F}_{\zeta_{i1,k}} \boldsymbol{\xi} = b_{\zeta_{i1,k}} + e_{\zeta_{i1,k}} \tag{8.40}$$

where

$$\boldsymbol{F}_{\zeta_{i1,k}} = \left[\cos\tilde{\theta}_{i,k} - \cos\tilde{\theta}_{1,k},\ \sin\tilde{\theta}_{i,k} - \sin\tilde{\theta}_{1,k}\right] \boldsymbol{L}, \tag{8.41a}$$

$$b_{\zeta_{i1,k}} = \tilde{\zeta}_{i1,k}, \tag{8.41b}$$

$$e_{\zeta_{i1,k}} \approx \left(v_{y,k}\cos\theta_{i,k} - v_{x,k}\sin\theta_{i,k}\right) n_{i,k} - \left(v_{y,k}\cos\theta_{1,k} - v_{x,k}\sin\theta_{1,k}\right) n_{1,k} - \varrho_{i1,k}. \tag{8.41c}$$

The expression for $e_{\zeta_{i1,k}}$ in (8.41c) neglects the second and higher-order noise terms under the small noise approximation. An exact expression for $e_{\zeta_{i1,k}}$ is derived in Section 8.11 (Appendix B).

8.4.2 Pseudolinear least-squares

Stacking (8.28), (8.37) and (8.40) for N AOA measurements, $N-1$ TDOA measurements and $N-1$ FDOA measurements at time instant k, we have

$$\boldsymbol{F}_k \boldsymbol{\xi} = \boldsymbol{b}_k + \boldsymbol{e}_k \tag{8.42}$$

where

$$\boldsymbol{F}_k = \left[\boldsymbol{F}_{\theta_{1,k}}^T, \ldots, \boldsymbol{F}_{\theta_{N,k}}^T, \boldsymbol{F}_{\gamma_{21,k}}^T, \ldots, \boldsymbol{F}_{\gamma_{N1,k}}^T, \boldsymbol{F}_{\zeta_{21,k}}^T, \ldots, \boldsymbol{F}_{\zeta_{N1,k}}^T\right]^T, \tag{8.43a}$$

$$\boldsymbol{b}_k = \left[b_{\theta_{1,k}}, \ldots, b_{\theta_{N,k}}, b_{\gamma_{21,k}}, \ldots, b_{\gamma_{N1,k}}, b_{\zeta_{21,k}}, \ldots, b_{\zeta_{N1,k}}\right]^T, \tag{8.43b}$$

$$\boldsymbol{e}_k = \left[e_{\theta_{1,k}}, \ldots, e_{\theta_{N,k}}, e_{\gamma_{21,k}}, \ldots, e_{\gamma_{N1,k}}, e_{\zeta_{21,k}}, \ldots, e_{\zeta_{N1,k}}\right]^T. \tag{8.43c}$$

Now stacking (8.42) for $k = 0, \ldots, M-1$ yields

$$\boldsymbol{F}\boldsymbol{\xi} = \boldsymbol{b} + \boldsymbol{e} \tag{8.44}$$

where

$$\boldsymbol{F} = \left[\boldsymbol{F}_0^T, \boldsymbol{F}_1^T, \ldots, \boldsymbol{F}_{M-1}^T\right]^T, \tag{8.45a}$$

$$\boldsymbol{b} = \left[\boldsymbol{b}_0^T, \boldsymbol{b}_1^T, \ldots, \boldsymbol{b}_{M-1}^T\right]^T, \tag{8.45b}$$

$$\boldsymbol{e} = \left[\boldsymbol{e}_0^T, \boldsymbol{e}_1^T, \ldots, \boldsymbol{e}_{M-1}^T\right]^T. \tag{8.45c}$$

Solving (8.44) in the least-squares sense results in the following pseudolinear estimator:

$$\hat{\boldsymbol{\xi}}_{\text{PLE}} = \arg\min_{\boldsymbol{\xi}\in\mathbb{R}^4} \|\boldsymbol{F}\boldsymbol{\xi} - \boldsymbol{b}\|^2 \tag{8.46a}$$

$$= (\boldsymbol{F}^T\boldsymbol{F})^{-1}\boldsymbol{F}^T\boldsymbol{b}. \tag{8.46b}$$

The target is observable if $(\boldsymbol{F}^T\boldsymbol{F})^{-1}$ exists (i.e., $\boldsymbol{F}$ is full-rank). For the problem under consideration, $\boldsymbol{F}$ is generally full-rank because the radar measurements are taken at a number of spatially distributed receivers during the target motion. The matrix $\boldsymbol{F}$ will be rank-deficient only for the special geometry (rarely occurring in practice) in which the target is collinear with all the receivers during the entire observation period.

8.5 Bias compensation for pseudolinear estimator

8.5.1 Bias analysis

The PLE estimate $\hat{\boldsymbol{\xi}}_{\text{PLE}}$ can be written in terms of the pseudolinear noise vector $\boldsymbol{e}$ by substituting $\boldsymbol{b} = \boldsymbol{F}\boldsymbol{\xi} - \boldsymbol{e}$ (see (8.44)) into (8.46b)

$$\hat{\boldsymbol{\xi}}_{\text{PLE}} = \boldsymbol{\xi} - (\boldsymbol{F}^T\boldsymbol{F})^{-1}\boldsymbol{F}^T\boldsymbol{e}. \tag{8.47}$$

Thus, the bias of the PLE is given by

$$\boldsymbol{\delta} = E\{\hat{\boldsymbol{\xi}}_{\text{PLE}}\} - \boldsymbol{\xi} = -E\left\{\left(\boldsymbol{F}^T\boldsymbol{F}\right)^{-1}\boldsymbol{F}^T\boldsymbol{e}\right\}. \tag{8.48}$$

As follows from Slutsky's theorem [100], the PLE bias can be approximated by

$$\boldsymbol{\delta} \approx -E\left\{\frac{\boldsymbol{F}^T\boldsymbol{F}}{M}\right\}^{-1} E\left\{\frac{\boldsymbol{F}^T\boldsymbol{e}}{M}\right\} \tag{8.49}$$

for sufficiently large M. Since the measurement matrix $\boldsymbol{F}$ is constructed from noisy AOA, TDOA and FDOA measurements, $\boldsymbol{F}$ is correlated with the pseudolinear noise vector $\boldsymbol{e}$ and this correlation does not vanish with increasing M, leading to

$$E\left\{\frac{\boldsymbol{F}^T\boldsymbol{e}}{M}\right\} \neq \boldsymbol{0}. \tag{8.50}$$

This consequently results in a nonzero bias for the PLE:

$$\boldsymbol{\delta}_{\text{PLE}} \neq \boldsymbol{0}. \tag{8.51}$$

Eq. (8.50) can be expressed as

$$E\left\{\frac{\boldsymbol{F}^T\boldsymbol{e}}{M}\right\} = \frac{1}{M}\sum_{k=0}^{M-1} E\{\boldsymbol{F}_k^T\boldsymbol{e}_k\} \tag{8.52a}$$

$$= \frac{1}{M}\sum_{k=0}^{M-1}\left(\sum_{i=1}^{N} E\left\{\boldsymbol{F}_{\theta_{i,k}}^T e_{\theta_{i,k}}\right\} + \sum_{i=2}^{N} E\left\{\boldsymbol{F}_{\gamma_{i1,k}}^T e_{\gamma_{i1,k}}\right\} + \sum_{i=2}^{N} E\left\{\boldsymbol{F}_{\zeta_{i1,k}}^T e_{\zeta_{i1,k}}\right\}\right) \tag{8.52b}$$

where $E\{\boldsymbol{F}_{\theta_{i,k}}^T e_{\theta_{i,k}}\}$, $E\{\boldsymbol{F}_{\gamma_{i1,k}}^T e_{\gamma_{i1,k}}\}$ and $E\{\boldsymbol{F}_{\zeta_{i1,k}}^T e_{\zeta_{i1,k}}\}$ are approximately written as

$$E\left\{\boldsymbol{F}_{\theta_{i,k}}^T e_{\theta_{i,k}}\right\} \approx \sigma_{\theta_{i,k}}^2 \boldsymbol{M}_k^T \boldsymbol{d}_{i,k}, \tag{8.53}$$

$$\begin{aligned}
E\left\{\boldsymbol{F}_{\gamma_{i1,k}}^T e_{\gamma_{i1,k}}\right\} \approx \frac{\boldsymbol{M}_k^T\boldsymbol{r}_{i1}}{\|\boldsymbol{r}_{i1}\|^2}\Bigg\{ & \frac{\|\boldsymbol{d}_{1,k}\|\cos(\theta_{i,k}-\theta_{i1})}{\|\boldsymbol{r}_{i1}\|}\boldsymbol{K}_{\boldsymbol{\varepsilon\varepsilon},k}(i-1) \\
& + \|\boldsymbol{d}_{i,k}\|(\|\boldsymbol{d}_{i,k}\| - \|\boldsymbol{d}_{1,k}\|)\sin^2(\theta_{i,k}-\theta_{i1})\sigma_{\theta_{i,k}}^2 \\
& + \left(\frac{\|\boldsymbol{d}_{i,k}\| - \|\boldsymbol{d}_{1,k}\|}{\|\boldsymbol{r}_{i1}\|} - \cos(\theta_{i,k}-\theta_{i1})\right) \\
& \times \bigg(-\frac{1}{2}\cos(\theta_{i,k}-\theta_{i1})\boldsymbol{K}_{\boldsymbol{\varepsilon\varepsilon},k}(i-1) \\
& \quad + \frac{1}{2}\|\boldsymbol{d}_{i,k}\|(\|\boldsymbol{d}_{i,k}\| - \|\boldsymbol{d}_{1,k}\|)\cos(\theta_{i,k}-\theta_{i1})\sigma_{\theta_{i,k}}^2\bigg)\Bigg\},
\end{aligned} \tag{8.54}$$

$$\begin{aligned}
& E\left\{\boldsymbol{F}_{\zeta_{i1,k}}^T e_{\zeta_{i1,k}}\right\} \\
& \approx \boldsymbol{L}^T\Bigg\{\begin{bmatrix} -\sin\theta_{i,k} \\ \cos\theta_{i,k} \end{bmatrix}(v_{y,k}\cos\theta_{i,k} - v_{x,k}\sin\theta_{i,k})\sigma_{\theta_{i,k}}^2 \\
& \quad + \begin{bmatrix} -\sin\theta_{1,k} \\ \cos\theta_{1,k} \end{bmatrix}(v_{y,k}\cos\theta_{1,k} - v_{x,k}\sin\theta_{1,k})\sigma_{\theta_{1,k}}^2
\end{aligned}$$

$$+\begin{bmatrix} \cos\theta_{i,k} - \cos\theta_{1,k} \\ \sin\theta_{i,k} - \sin\theta_{1,k} \end{bmatrix} \times \left(-\frac{(v_{x,k}\cos\theta_{i,k} + v_{y,k}\sin\theta_{i,k})}{2}\sigma^2_{\theta_{i,k}} + \frac{(v_{x,k}\cos\theta_{1,k} + v_{y,k}\sin\theta_{1,k})}{2}\sigma^2_{\theta_{1,k}} \right) \Bigg\},$$

(8.55)

under the small noise assumption. Here, $\boldsymbol{K}_{\boldsymbol{\varepsilon}\boldsymbol{\varepsilon},k}(i-1)$ denotes the $(i-1)$th diagonal entry of $\boldsymbol{K}_{\boldsymbol{\varepsilon}\boldsymbol{\varepsilon},k}$.

8.5.2 Bias compensation

A direct approach to moderating the bias problem of the PLE is to estimate and subtract the instantaneous PLE bias from the PLE estimate. For sufficiently large M, the instantaneous PLE bias can be estimated as

$$\hat{\boldsymbol{\xi}}_{\text{PLE}} - \boldsymbol{\xi} = -(\boldsymbol{F}^T\boldsymbol{F})^{-1}\boldsymbol{F}^T\boldsymbol{e} \tag{8.56a}$$

$$\approx -(\boldsymbol{F}^T\boldsymbol{F})^{-1}E\left\{\boldsymbol{F}^T\boldsymbol{e}\right\} \tag{8.56b}$$

$$\approx -(\boldsymbol{F}^T\boldsymbol{F})^{-1}\hat{E}\left\{\boldsymbol{F}^T\boldsymbol{e}\right\} \tag{8.56c}$$

where

$$\hat{E}\left\{\boldsymbol{F}^T\boldsymbol{e}\right\} = \sum_{k=0}^{M-1}\left(\sum_{i=1}^{N}\hat{E}\left\{\boldsymbol{F}^T_{\theta_{i,k}}e_{\theta_{i,k}}\right\} + \sum_{i=2}^{N}\hat{E}\left\{\boldsymbol{F}^T_{\gamma_{i1,k}}e_{\gamma_{i1,k}}\right\} + \sum_{i=2}^{N}\hat{E}\left\{\boldsymbol{F}^T_{\zeta_{i1,k}}e_{\zeta_{i1,k}}\right\}\right). \tag{8.57}$$

The expressions of $\hat{E}\{\boldsymbol{F}^T_{\theta_{i,k}}e_{\theta_{i,k}}\}$, $\hat{E}\{\boldsymbol{F}^T_{\gamma_{i1,k}}e_{\gamma_{i1,k}}\}$ and $\hat{E}\{\boldsymbol{F}^T_{\zeta_{i1,k}}e_{\zeta_{i1,k}}\}$ are identical to those of $E\{\boldsymbol{F}^T_{\theta_{i,k}}e_{\theta_{i,k}}\}$, $E\{\boldsymbol{F}^T_{\gamma_{i1,k}}e_{\gamma_{i1,k}}\}$ and $E\{\boldsymbol{F}^T_{\zeta_{i1,k}}e_{\zeta_{i1,k}}\}$ in (8.53)–(8.55), respectively; except that $\boldsymbol{d}_{i,k}$, $\theta_{i,k}$ and $\boldsymbol{v}_k$ are replaced by $\hat{\boldsymbol{d}}_{i,k}$, $\hat{\theta}_{i,k}$ and $\hat{\boldsymbol{v}}_k$ which are computed from the PLE estimate $\hat{\boldsymbol{\xi}}_{\text{PLE}}$. By subtracting the instantaneous PLE bias estimate given in (8.56) from $\hat{\boldsymbol{\xi}}_{\text{PLE}}$, we obtain a bias-compensated variant of the PLE:

$$\hat{\boldsymbol{\xi}}_{\text{BCPLE}} = \hat{\boldsymbol{\xi}}_{\text{PLE}} + (\boldsymbol{F}^T\boldsymbol{F})^{-1}\hat{E}\left\{\boldsymbol{F}^T\boldsymbol{e}\right\}. \tag{8.58}$$

The BCPLE is not strictly unbiased as the instantaneous PLE bias is estimated from the noisy measurements. Nevertheless, it is capable of significantly reducing the PLE bias, especially in the presence of small measurement noise [80].

8.6 Asymptotically-unbiased weighted instrumental variable estimator

An attractive alternative to the direct bias compensation method presented in the previous section is to exploit the use of instrumental variables. Specifically, the correlation between the measurement matrix $\boldsymbol{F}$ and the pseudolinear noise vector $\boldsymbol{e}$, which is the main cause of the PLE bias, can be almost eliminated by replacing $\boldsymbol{F}^T$ with an IV matrix $\boldsymbol{G}^T$ in the PLE normal equations, where $\boldsymbol{G}$ is approximately uncorrelated with the pseudolinear noise vector $\boldsymbol{e}$. Specifically, the normal equations of the PLE can be modified from

$$\boldsymbol{F}^T\boldsymbol{F}\hat{\boldsymbol{\xi}}_{\text{PLE}} = \boldsymbol{F}^T\boldsymbol{b} \tag{8.59}$$

to

$$\boldsymbol{G}^T\boldsymbol{F}\hat{\boldsymbol{\xi}}_{\text{IVE}} = \boldsymbol{G}^T\boldsymbol{b} \tag{8.60}$$

which leads to the following IV estimator (IVE):

$$\hat{\boldsymbol{\xi}}_{\text{IVE}} = (\boldsymbol{G}^T\boldsymbol{F})^{-1}\boldsymbol{G}^T\boldsymbol{b}. \tag{8.61}$$

To ensure the asymptotic unbiasedness of the IVE, i.e., $E\{\hat{\boldsymbol{\xi}}_{\text{IVE}}\} = \boldsymbol{\xi}$ as $M \to \infty$, the IV matrix $\boldsymbol{G}$ must be selected such that $E\{\boldsymbol{G}^T\boldsymbol{F}/M\}$ is nonsingular and $E\{\boldsymbol{G}^T\boldsymbol{e}/M\} = \boldsymbol{0}$ as $M \to \infty$.

In general, the noise-free version of the measurement matrix $\boldsymbol{F}$ (denoted as $\boldsymbol{F}_\circ$) is the optimal choice for the IV matrix. However, since $\boldsymbol{F}_\circ$ is unknown, we utilize an estimate of $\boldsymbol{F}_\circ$ calculated from the BCPLE estimate $\hat{\boldsymbol{\xi}}_{\text{BCPLE}}$ as the IV matrix. Specifically, the IV matrix $\boldsymbol{G}$ is given by

$$\boldsymbol{G} = \left[\boldsymbol{G}_0^T, \boldsymbol{G}_1^T, \ldots, \boldsymbol{G}_{M-1}^T\right]^T \tag{8.62}$$

where

$$\boldsymbol{G}_k = \left[\boldsymbol{G}_{\theta_{1,k}}^T, \ldots, \boldsymbol{G}_{\theta_{N,k}}^T, \boldsymbol{G}_{\gamma_{21,k}}^T \ldots, \boldsymbol{G}_{\gamma_{N1,k}}^T, \boldsymbol{G}_{\zeta_{21,k}}^T, \ldots, \boldsymbol{G}_{\zeta_{N1,k}}^T\right]^T \tag{8.63}$$

and

$$\boldsymbol{G}_{\theta_{i,k}} = \left[\sin\hat{\theta}_{i,k},\ \cos\hat{\theta}_{i,k}\right]\boldsymbol{M}_k, \tag{8.64a}$$

$$\boldsymbol{G}_{\gamma_{i1,k}} = \left(\frac{\hat{\gamma}_{i1,k}}{\|\boldsymbol{r}_{i1}\|} - \cos(\hat{\theta}_{i,k} - \theta_{i1})\right)\frac{\boldsymbol{r}_{i1}^T\boldsymbol{M}_k}{\|\boldsymbol{r}_{i1}\|}, \tag{8.64b}$$

$$\boldsymbol{G}_{\zeta_{i1,k}} = \left[\cos\hat{\theta}_{i,k} - \cos\hat{\theta}_{1,k},\ \sin\hat{\theta}_{i,k} - \sin\hat{\theta}_{1,k}\right]\boldsymbol{L}, \tag{8.64c}$$

with

$$\hat{\theta}_{i,k} = \tan^{-1}\frac{\hat{p}_{y,k} - r_{y,i}}{\hat{p}_{x,k} - r_{x,i}}, \tag{8.65a}$$

$$\hat{\gamma}_{ji,k} = \|\hat{\boldsymbol{p}}_k - \boldsymbol{r}_j\| - \|\hat{\boldsymbol{p}}_k - \boldsymbol{r}_i\|, \tag{8.65b}$$

$$\hat{\boldsymbol{p}}_k = [\hat{p}_{x,k}, \hat{p}_{y,k}]^T = \boldsymbol{M}_k\hat{\boldsymbol{\xi}}_{\text{BCPLE}}. \tag{8.65c}$$

Since the covariance of $\hat{\boldsymbol{\xi}}_{\text{BCPLE}}$ tends to zero as $M \to \infty$, the correlation between $\hat{\boldsymbol{\xi}}_{\text{BCPLE}}$ and $\boldsymbol{e}$ vanishes as $M \to \infty$. Therefore, $\mathbf{G}$ constructed using $\hat{\boldsymbol{\xi}}_{\text{BCPLE}}$ is uncorrelated asymptotically with $\boldsymbol{e}$ [80], i.e.,

$$E\left\{\frac{\boldsymbol{G}^T\boldsymbol{e}}{M}\right\} = \boldsymbol{0}, \quad \text{as } M \to \infty. \tag{8.66}$$

In practice, the approximation of $E\{\boldsymbol{G}^T\boldsymbol{e}/M\} \approx \boldsymbol{0}$ holds for sufficiently large M.

In order to improve the RMSE performance of the IVE, we introduce a weighting matrix that approximates the covariance of the pseudolinear noise vector $\boldsymbol{e}$:

$$\boldsymbol{W}_\circ = E\{\boldsymbol{e}\boldsymbol{e}^T\} = \text{diag}\left(\boldsymbol{W}_0, \boldsymbol{W}_1, \ldots, \boldsymbol{W}_{M-1}\right) \tag{8.67}$$

where

$$\boldsymbol{W}_k = E\{\boldsymbol{e}_k\boldsymbol{e}_k^T\} = \begin{bmatrix} W_{11,k} & \boldsymbol{W}_{12,k} & \boldsymbol{W}_{13,k} & \boldsymbol{W}_{14,k} \\ \boldsymbol{W}_{12,k}^T & \boldsymbol{W}_{22,k} & \boldsymbol{W}_{23,k} & \boldsymbol{W}_{24,k} \\ \boldsymbol{W}_{13,k}^T & \boldsymbol{W}_{23,k}^T & \boldsymbol{W}_{33,k} & \boldsymbol{W}_{34,k} \\ \boldsymbol{W}_{14,k}^T & \boldsymbol{W}_{24,k}^T & \boldsymbol{W}_{34,k}^T & \boldsymbol{W}_{44,k} \end{bmatrix}. \tag{8.68}$$

The entries of $\boldsymbol{W}_k$ are given by

$$W_{11,k} = E\{e_{\theta_{1,k}}^2\} \approx \|\boldsymbol{d}_{1,k}\|^2\sigma_{\theta_{1,k}}^2, \tag{8.69a}$$

$$\boldsymbol{W}_{12,k} = E\left\{e_{\theta_{1,k}}[e_{\theta_{2,k}}, \ldots, e_{\theta_{N,k}}]\right\} = \boldsymbol{0}_{1\times(N-1)}, \tag{8.69b}$$

$$\boldsymbol{W}_{13,k} = E\left\{e_{\theta_{1,k}}[e_{\gamma_{21,k}}, \ldots, e_{\gamma_{N1,k}}]\right\} = \boldsymbol{0}_{1\times(N-1)}, \tag{8.69c}$$

$$\begin{aligned}\boldsymbol{W}_{14,k} =& E\left\{e_{\theta_{1,k}}[e_{\zeta_{21,k}}, \ldots, e_{\zeta_{N1,k}}]\right\} \\ &\approx -\|\boldsymbol{d}_{1,k}\|\left(v_{y,k}\cos\theta_{1,k} - v_{x,k}\sin\theta_{1,k}\right)\sigma_{\theta_{1,k}}^2 \times \boldsymbol{1}_{1\times(N-1)},\end{aligned} \tag{8.69d}$$

$$\begin{aligned}\boldsymbol{W}_{22,k} =& E\left\{[e_{\theta_{2,k}}, \ldots, e_{\theta_{N,k}}]^T[e_{\theta_{2,k}}, \ldots, e_{\theta_{N,k}}]\right\} \\ &\approx \text{diag}\left(\ldots, \|\boldsymbol{d}_{i,k}\|^2\sigma_{\theta_{i,k}}^2, \ldots\right)_{i=2,\ldots,N},\end{aligned} \tag{8.69e}$$

$$\begin{aligned} \boldsymbol{W}_{23,k} &= E\left\{[e_{\theta_{2,k}}, \ldots, e_{\theta_{N,k}}]^T [e_{\gamma_{21,k}}, \ldots, e_{\gamma_{N1,k}}]\right\} \\ &\approx \operatorname{diag}\left(\ldots, \frac{\|\boldsymbol{d}_{i,k}\|^2 \left(\|\boldsymbol{d}_{i,k}\| - \|\boldsymbol{d}_{1,k}\|\right)}{\|\boldsymbol{r}_{i1}\|} \sin(\theta_{i,k} - \theta_{i1})\sigma^2_{\theta_{i,k}}, \ldots\right)_{i=2,\ldots,N}, \end{aligned} \tag{8.69f}$$

$$\begin{aligned} \boldsymbol{W}_{24,k} &= E\left\{[e_{\theta_{2,k}}, \ldots, e_{\theta_{N,k}}]^T [e_{\zeta_{21,k}}, \ldots, e_{\zeta_{N1,k}}]\right\} \\ &\approx \operatorname{diag}\left(\ldots, \|\boldsymbol{d}_{i,k}\| \left(v_{y,k}\cos\theta_{i,k} - v_{x,k}\sin\theta_{i,k}\right)\sigma^2_{\theta_{i,k}}, \ldots\right)_{i=2,\ldots,N}, \end{aligned} \tag{8.69g}$$

$$\begin{aligned} \boldsymbol{W}_{33,k} &= E\left\{[e_{\gamma_{21,k}}, \ldots, e_{\gamma_{N1,k}}]^T [e_{\gamma_{21,k}}, \ldots, e_{\gamma_{N1,k}}]\right\} \\ &\approx \operatorname{diag}\left(\ldots, \frac{\|\boldsymbol{d}_{1,k}\|}{\|\boldsymbol{r}_{i1}\|}\cos(\theta_{i,k} - \theta_{i1}), \ldots\right)_{i=2,\ldots,N} \\ &\quad \times \boldsymbol{K}_{\boldsymbol{\varepsilon\varepsilon},k} \times \operatorname{diag}\left(\ldots, \frac{\|\boldsymbol{d}_{1,k}\|}{\|\boldsymbol{r}_{i1}\|}\cos(\theta_{i,k} - \theta_{i1}), \ldots\right)_{i=2,\ldots,N} \\ &\quad + \operatorname{diag}\left(\ldots, \frac{\|\boldsymbol{d}_{i,k}\|^2 \left(\|\boldsymbol{d}_{i,k}\| - \|\boldsymbol{d}_{1,k}\|\right)^2}{\|\boldsymbol{r}_{i1}\|^2} \right. \\ &\qquad \left. \times \sin^2(\theta_{i,k} - \theta_{i1})\sigma^2_{\theta_{i,k}}, \ldots\right)_{i=2,\ldots,N}, \end{aligned} \tag{8.69h}$$

$$\begin{aligned} \boldsymbol{W}_{34,k} &= E\left\{[e_{\gamma_{21,k}}, \ldots, e_{\gamma_{N1,k}}]^T [e_{\zeta_{21,k}}, \ldots, e_{\zeta_{N1,k}}]\right\} \\ &\approx -\operatorname{diag}\left(\ldots, \frac{\|\boldsymbol{d}_{1,k}\|}{\|\boldsymbol{r}_{i1}\|}\cos(\theta_{i,k} - \theta_{i1}), \ldots\right)_{i=2,\ldots,N} \times \boldsymbol{K}_{\boldsymbol{\varepsilon\varrho},k} \\ &\quad + \operatorname{diag}\left(\ldots, \frac{\|\boldsymbol{d}_{i,k}\| \left(\|\boldsymbol{d}_{i,k}\| - \|\boldsymbol{d}_{1,k}\|\right)}{\|\boldsymbol{r}_{i1}\|} \right. \\ &\qquad \left. \times \sin(\theta_{i,k} - \theta_{i1})\left(v_{y,k}\cos\theta_{i,k} - v_{x,k}\sin\theta_{i,k}\right)\sigma^2_{\theta_{i,k}}, \ldots\right)_{i=2,\ldots,N}, \end{aligned} \tag{8.69i}$$

$$\begin{aligned} \boldsymbol{W}_{44,k} &= E\left\{[e_{\zeta_{21,k}}, \ldots, e_{\zeta_{N1,k}}]^T [e_{\zeta_{21,k}}, \ldots, e_{\zeta_{N1,k}}]\right\} \\ &\approx \operatorname{diag}\left(\ldots, (v_{y,k}\cos\theta_{i,k} - v_{x,k}\sin\theta_{i,k})^2\sigma^2_{\theta_{i,k}}, \ldots\right)_{i=2,\ldots,N} \\ &\quad + (v_{y,k}\cos\theta_{1,k} - v_{x,k}\sin\theta_{1,k})^2\sigma^2_{\theta_{1,k}} \times \mathbf{1}_{(N-1)\times(N-1)} + \boldsymbol{K}_{\boldsymbol{\varrho\varrho},k}, \end{aligned} \tag{8.69j}$$

where the third- and higher-order noise terms are neglected under the assumption of small measurement noise. It is important to note that there is a correlation between the pseudolinear noise of AOA, TDOA and FDOA components because AOA measurements are used jointly with TDOA and FDOA measurements to formulate the pseudolinear equations for TDOA and FDOA (see Section 8.4). This is reflected by the non-diagonal form of $\boldsymbol{W}_k$. The entries of $\boldsymbol{W}_k$ in (8.69) are functions of the unknown terms $\boldsymbol{d}_{i,k}$, $\theta_{i,k}$ and $\boldsymbol{v}_k$, thus $\boldsymbol{W}_\circ$ in (8.67) cannot be computed. Therefore, the estimates $\hat{\boldsymbol{d}}_{i,k}$, $\hat{\theta}_{i,k}$ and $\hat{\boldsymbol{v}}_k$ obtained from the BCPLE estimate $\hat{\boldsymbol{\xi}}_{\text{BCPLE}}$ are utilized instead to compute the weighting matrix $\boldsymbol{W}$. Note that we will refer to $\boldsymbol{W}_\circ$ as the noise-free version of $\boldsymbol{W}$.

Applying the weighting matrix $\boldsymbol{W}$ to the IVE in (8.61) yields the WIVE

$$\hat{\boldsymbol{\xi}}_{\text{WIVE}} = (\boldsymbol{G}^T\boldsymbol{W}^{-1}\boldsymbol{F})^{-1}\boldsymbol{G}^T\boldsymbol{W}^{-1}\boldsymbol{b}. \tag{8.70}$$

8.7 Asymptotic efficiency analysis

In this section, the WIVE is analytically shown to be asymptotically efficient (i.e., its error covariance matrix approaches the CRLB matrix as $M \to \infty$) in the presence of small measurement noise.

The error covariance matrix of the WIVE is given by

$$\begin{aligned} \boldsymbol{C}_{\text{WIVE}} &= E\left\{(\hat{\boldsymbol{\xi}}_{\text{WIVE}} - \boldsymbol{\xi})(\hat{\boldsymbol{\xi}}_{\text{WIVE}} - \boldsymbol{\xi})^T\right\} & \text{(8.71a)} \\ &= E\left\{\left(\boldsymbol{G}^T\boldsymbol{W}^{-1}\boldsymbol{F}\right)^{-1}\boldsymbol{G}^T\boldsymbol{W}^{-1}\boldsymbol{e}\boldsymbol{e}^T\boldsymbol{W}^{-1}\boldsymbol{G}\left(\boldsymbol{F}^T\boldsymbol{W}^{-1}\boldsymbol{G}\right)^{-1}\right\}. & \text{(8.71b)} \end{aligned}$$

Assuming the observability of the target (i.e., $\boldsymbol{G}^T\boldsymbol{W}^{-1}\boldsymbol{F}$ is nonsingular), the asymptotic error covariance of the WIVE is given by

$$\boldsymbol{C}_{\text{WIVE}} = \text{plim}\left\{\frac{1}{M}\left(\frac{\boldsymbol{G}^T\boldsymbol{W}^{-1}\boldsymbol{F}}{M}\right)^{-1}\frac{\boldsymbol{G}^T\boldsymbol{W}^{-1}\boldsymbol{e}\boldsymbol{e}^T\boldsymbol{W}^{-1}\boldsymbol{G}}{M}\left(\frac{\boldsymbol{F}^T\boldsymbol{W}^{-1}\boldsymbol{G}}{M}\right)^{-1}\right\} \tag{8.72}$$

where the probability limit (plim) is defined by [101]

$$\text{plim}\ \hat{\boldsymbol{x}}(M) = \boldsymbol{x}^* \iff \lim_{M\to\infty} P\{|\boldsymbol{x}(M) - \boldsymbol{x}^*| > \epsilon\} = 0$$

for every $\epsilon > 0$. Applying Slutsky's theorem [100] to (8.72), we obtain

$$\begin{aligned} \boldsymbol{C}_{\text{WIVE}} = \text{plim}\left\{\frac{1}{M}\right\}\text{plim}\left\{\frac{\boldsymbol{G}^T\boldsymbol{W}^{-1}\boldsymbol{F}}{M}\right\}^{-1}\text{plim}\left\{\frac{\boldsymbol{G}^T\boldsymbol{W}^{-1}\boldsymbol{e}\boldsymbol{e}^T\boldsymbol{W}^{-1}\boldsymbol{G}}{M}\right\} \\ \times\ \text{plim}\left\{\frac{\boldsymbol{F}^T\boldsymbol{W}^{-1}\boldsymbol{G}}{M}\right\}^{-1}. \end{aligned} \tag{8.73}$$

We can rewrite the term inside the second plim in the right-hand side of (8.73) as

$$\frac{\boldsymbol{G}^T\boldsymbol{W}^{-1}\boldsymbol{F}}{M}=\frac{1}{M}\sum_{k=0}^{M-1}\boldsymbol{G}_k^T\boldsymbol{W}_k^{-1}\boldsymbol{F}_k. \tag{8.74}$$

Since the measurement noise vector $\boldsymbol{\eta}_k$ is finite and statistically independent over k, $\boldsymbol{G}_k^T\boldsymbol{W}_k^{-1}\boldsymbol{F}_k$ is also statistically independent over k and the entries of $\boldsymbol{G}_k^T\boldsymbol{W}_k^{-1}\boldsymbol{F}_k$ (denoted as $\kappa_{ij,k}$) have finite variances. Using the inequality

$$\sum_{k=0}^{M-1}\frac{E\left\{\kappa_{ij,k}^2\right\}}{(k+1)^2}\le\max\left\{E\left\{\kappa_{ij,k}^2\right\}\right\}\sum_{k=0}^{M-1}\frac{1}{(k+1)^2} \tag{8.75}$$

and noting that $\sum_{k=0}^{M-1}\frac{1}{(k+1)^2}$ converges to $\pi^2/6$ as $M\to\infty$ [102], $\sum_{k=0}^{M-1}\frac{E\left\{\kappa_{ij,k}^2\right\}}{(k+1)^2}$ converges as $M\to\infty$. Therefore, the Kolmogorov criterion is satisfied for each entry $\kappa_{ij,k}$ of $\boldsymbol{G}_k^T\boldsymbol{W}_k^{-1}\boldsymbol{F}_k$. Consequently, following Kolmogorov's strong law of large numbers [103], $\frac{1}{M}\sum_{k=0}^{M-1}\boldsymbol{G}_k^T\boldsymbol{W}_k^{-1}\boldsymbol{F}_k$ almost surely converges to $\frac{1}{M}\sum_{k=0}^{M-1}E\{\boldsymbol{G}_k^T\boldsymbol{W}_k^{-1}\boldsymbol{F}_k\}$ as $M\to\infty$. This implies that the sample covariance matrix $\frac{\boldsymbol{G}^T\boldsymbol{W}^{-1}\boldsymbol{F}}{M}$ almost surely converges to $E\left\{\frac{\boldsymbol{G}^T\boldsymbol{W}^{-1}\boldsymbol{F}}{M}\right\}$ as $M\to\infty$. Therefore, we obtain

$$\text{plim}\left\{\frac{\boldsymbol{G}^T\boldsymbol{W}^{-1}\boldsymbol{F}}{M}\right\}^{-1}=E\left\{\frac{\boldsymbol{G}^T\boldsymbol{W}^{-1}\boldsymbol{F}}{M}\right\}^{-1}. \tag{8.76}$$

For small measurement noise, the BCPLE estimate $\hat{\boldsymbol{\xi}}_{\text{BCPLE}}$, which is used to compute $\boldsymbol{G}$ and $\boldsymbol{W}$ for the WIVE, has a small bias. In addition, the covariances of the estimates $\hat{\boldsymbol{d}}_{i,k}$, $\hat{\theta}_{i,k}$ and $\hat{\boldsymbol{v}}_k$ computed from $\hat{\boldsymbol{\xi}}_{\text{BCPLE}}$ vanish as $M\to\infty$. Therefore, $\boldsymbol{G}$ and $\boldsymbol{W}$ can be approximated by their noise-free versions $\boldsymbol{G}_\circ$ and $\boldsymbol{W}_\circ$ as $M\to\infty$ under the small measurement noise assumption. Note that $\boldsymbol{G}_\circ=\boldsymbol{F}_\circ$. Consequently, we have

$$\text{plim}\left\{\frac{\boldsymbol{G}^T\boldsymbol{W}^{-1}\boldsymbol{F}}{M}\right\}^{-1}\approx E\left\{\frac{\boldsymbol{F}_\circ^T\boldsymbol{W}_\circ^{-1}\boldsymbol{F}}{M}\right\}^{-1}. \tag{8.77}$$

Using the approximation $E\{\boldsymbol{F}\}\approx\boldsymbol{F}_\circ$ (by neglecting the second- and higher-order noise terms) leads to

$$E\left\{\frac{\boldsymbol{F}_\circ^T\boldsymbol{W}_\circ^{-1}\boldsymbol{F}}{M}\right\}=\frac{\boldsymbol{F}_\circ^T\boldsymbol{W}_\circ^{-1}E\{\boldsymbol{F}\}}{M}\approx\frac{\boldsymbol{F}_\circ^T\boldsymbol{W}_\circ^{-1}\boldsymbol{F}_\circ}{M}. \tag{8.78}$$

From (8.77) and (8.78), we obtain

$$\operatorname{plim}\left\{\frac{\boldsymbol{G}^T\boldsymbol{W}^{-1}\boldsymbol{F}}{M}\right\}^{-1} \approx M\left(\boldsymbol{F}_\circ^T\boldsymbol{W}_\circ^{-1}\boldsymbol{F}_\circ\right)^{-1}. \tag{8.79}$$

Similarly, we have

$$\operatorname{plim}\left\{\frac{\boldsymbol{F}^T\boldsymbol{W}^{-1}\boldsymbol{G}}{M}\right\}^{-1} \approx M\left(\boldsymbol{F}_\circ^T\boldsymbol{W}_\circ^{-1}\boldsymbol{F}_\circ\right)^{-1}. \tag{8.80}$$

Using a similar approach that leads to (8.76), we can use the strong law of large numbers to show that

$$\operatorname{plim}\left\{\frac{\boldsymbol{G}^T\boldsymbol{W}^{-1}\boldsymbol{e}\boldsymbol{e}^T\boldsymbol{W}^{-1}\boldsymbol{G}}{M}\right\} = E\left\{\frac{\boldsymbol{G}^T\boldsymbol{W}^{-1}\boldsymbol{e}\boldsymbol{e}^T\boldsymbol{W}^{-1}\boldsymbol{G}}{M}\right\}. \tag{8.81}$$

Using small noise approximations, we obtain

$$\begin{aligned}\operatorname{plim}\left\{\frac{\boldsymbol{G}^T\boldsymbol{W}^{-1}\boldsymbol{e}\boldsymbol{e}^T\boldsymbol{W}^{-1}\boldsymbol{G}}{M}\right\} &\approx \boldsymbol{F}_\circ^T\boldsymbol{W}_\circ^{-1}E\left\{\frac{\boldsymbol{e}\boldsymbol{e}^T}{M}\right\}\boldsymbol{W}_\circ^{-1}\boldsymbol{F}_\circ \\ &\approx \frac{\boldsymbol{F}_\circ^T\boldsymbol{W}_\circ^{-1}\boldsymbol{W}_\circ\boldsymbol{W}_\circ^{-1}\boldsymbol{F}_\circ}{M} \\ &\approx \frac{\boldsymbol{F}_\circ^T\boldsymbol{W}_\circ^{-1}\boldsymbol{F}_\circ}{M}.\end{aligned} \tag{8.82}$$

By substituting (8.79), (8.80) and (8.82) into (8.73), we obtain the asymptotic error covariance of the WIVE as $M \to \infty$:

$$\boldsymbol{C}_{\mathrm{WIVE}} \approx \left(\boldsymbol{F}_\circ^T\boldsymbol{W}_\circ^{-1}\boldsymbol{F}_\circ\right)^{-1}. \tag{8.83}$$

Since it is proved in Section 8.11 (Appendix A) that

$$\boldsymbol{F}_\circ^T\boldsymbol{W}_\circ^{-1}\boldsymbol{F}_\circ = \boldsymbol{J}_\circ^T\boldsymbol{K}^{-1}\boldsymbol{J}_\circ \tag{8.84}$$

we have

$$\boldsymbol{C}_{\mathrm{WIVE}} \approx \left(\boldsymbol{J}_\circ^T\boldsymbol{K}^{-1}\boldsymbol{J}_\circ\right)^{-1} \tag{8.85a}$$

$$\approx \boldsymbol{C}_{\boldsymbol{\xi}}. \tag{8.85b}$$

This concludes the proof for the asymptotic efficiency of the WIVE under the small measurement noise assumption.

Table 8.1 Comparison of computational complexities.

Algorithm	Multiplication	Division	Root	Inv. 1[†]	Inv. 2[‡]
PLE	$M(74N-50)$	–	–	–	1
BCPLE	$M(105N-65)$	–	MN	–	1
WIVE	$M(39N^2+137N-99)$	–	$2MN$	M	2
Gauss–Newton[#]	$M(36N^2+29N-24)i_{GN}+M(105N-65)$	MNi_{GN}	$MN(i_{GN}+1)$	Mi_{GN}	$i_{GN}+1$

† Inv. 1 stands for $(3N-2)\times(3N-2)$ matrix inversion.
‡ Inv. 2 stands for 4×4 matrix inversion.
# The Gauss–Newton algorithm is initialized to the BCPLE and halted after i_{GN} iterations. The Gauss–Newton complexity also includes the computation of its initialization.

It is also shown in Section 8.11 (Appendix C) that the CRLB for the TMA problem under consideration is independent of the choice of TDOA/FDOA reference receiver. As a result, the asymptotic performance of the WIVE is independent of the choice of reference receiver.

8.8 Computational complexity

Table 8.1 summarizes the computational complexities of the PLE, BCPLE, WIVE and the Gauss–Newton implementation of the MLE. The complexity analysis excludes quantities that can be precalculated including $1/\|\boldsymbol{r}_{i1}\|$, $-\|\boldsymbol{r}_{i1}\|^2/2$, $\boldsymbol{r}_{i1}^T\boldsymbol{M}_k/\|\boldsymbol{r}_{i1}\|$ and $\boldsymbol{r}_{i1}^T\boldsymbol{r}_i/\|\boldsymbol{r}_{i1}\|$. For the multiplication operation, only the complexity terms dependent on M are shown in Table 8.1 since other terms are negligible for large M. Table 8.1 confirms the computational effectiveness of the closed-form algorithms (i.e., PLE, BCPLE and WIVE) over the iterative MLE which is computationally intensive due to its iterative nature.

8.9 Algorithm performance and comparison

Fig. 8.2 depicts a simulated TMA scenario with four receivers located at $\boldsymbol{r}_1=[0,0]^T$, $\boldsymbol{r}_2=[300,200]^T$, $\boldsymbol{r}_3=[0,400]^T$ and $\boldsymbol{r}_4=[-100,300]^T$ m. The target travels with a constant velocity of $\boldsymbol{v}_0=[10,5]^T$ m/s. The initial target position is set to $\boldsymbol{p}_0=[300,500]^T$ m. The receivers take measurements at time $t_k=kT/(M-1)$, $k=0,\ldots,M-1$ with $T=14.5$ s and $M=30$. The measurement noise is assumed to have a constant covariance over time: $\boldsymbol{K}_0=\boldsymbol{K}_1=\cdots=\boldsymbol{K}_{M-1}$. The AOA noise variance is assumed to be identical

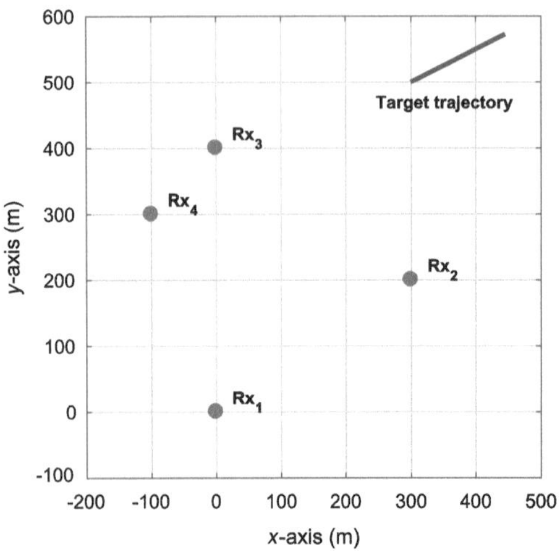

Figure 8.2 Simulated TMA geometry.

across different receivers $\sigma^2_{\theta_{1,k}} = \cdots = \sigma^2_{\theta_{N,k}} = \sigma^2_\theta$. In this simulation, we assume that the TDOA and FDOA noises are uncorrelated, i.e., $\boldsymbol{K}_{\boldsymbol{\varepsilon}\boldsymbol{\varrho},k} = \mathbf{0}$. In acoustic applications with stationary Gaussian signals, the TDOA noise is uncorrelated with the FDOA noise in the asymptotic sense and the CRLB matrix for joint TDOA/FDOA estimation has a block-diagonal form under relatively mild conditions when the signal time-bandwidth product is sufficiently large while the signal-to-noise ratio is moderately high [104–106]. Note that the algorithms presented in this chapter are also applicable to the case where the TDOA and FDOA noises are correlated such that $\boldsymbol{K}_{\boldsymbol{\varepsilon}\boldsymbol{\varrho},k} \neq \mathbf{0}$. For example, a strong correlation between the TDOA and FDOA noises exists in the case of deterministic non-stationary non-Gaussian signals, for example, chirp-like signals in radar applications [105,106]. Following [107], we model the TDOA noise covariance matrix $\boldsymbol{K}_{\boldsymbol{\varepsilon}\boldsymbol{\varepsilon},k}$ as

$$\boldsymbol{K}_{\boldsymbol{\varepsilon}\boldsymbol{\varepsilon},k} = \sigma^2_\varepsilon \boldsymbol{I}_{(M-1)\times(M-1)} + \sigma^2_\varepsilon \mathbf{1}_{(M-1)\times(M-1)} \tag{8.86}$$

and the FDOA noise covariance matrix $\boldsymbol{K}_{\boldsymbol{\varrho}\boldsymbol{\varrho},k}$ as

$$\boldsymbol{K}_{\boldsymbol{\varrho}\boldsymbol{\varrho},k} = \sigma^2_\varrho \boldsymbol{I}_{(M-1)\times(M-1)} + \sigma^2_\varrho \mathbf{1}_{(M-1)\times(M-1)}. \tag{8.87}$$

In the simulations, the Gauss–Newton algorithm is used to implement the MLE using an initial solution value of $\hat{\boldsymbol{\xi}}_{\mathrm{BCPLE}}$ (the BCPLE estimate). The Gauss–Newton algorithm is halted after 10 iterations.

Table 8.2 Noise levels used in Example 1.

Noise index	1	2	3	4	5	6	7
σ_θ (degree)	1	2	3	4	5	6	7
σ_ε (m)	2.5	5	7.5	10	12.5	15	17.5
σ_ϱ (m/s)	0.1	0.2	0.3	0.4	0.5	0.6	0.7

The bias norm, RMSE and CRLB are used for performance comparison. The bias norm and RMSE for position estimate are obtained via Monte Carlo simulations using

$$\text{Bias norm}_{\text{pos}} = \left\| \frac{1}{N_{\text{MC}}} \sum_{l=1}^{N_{\text{MC}}} \left(\hat{\boldsymbol{p}}_0^{\langle l \rangle} - \boldsymbol{p}_0 \right) \right\|, \tag{8.88a}$$

$$\text{RMSE}_{\text{pos}} = \left(\frac{1}{N_{\text{MC}}} \sum_{l=1}^{N_{\text{MC}}} \left\| \hat{\boldsymbol{p}}_0^{\langle l \rangle} - \boldsymbol{p}_0 \right\|^2 \right)^{1/2}, \tag{8.88b}$$

where $\hat{\boldsymbol{p}}_0^{\langle l \rangle} = [\hat{\boldsymbol{\xi}}^{\langle l \rangle}(1), \hat{\boldsymbol{\xi}}^{\langle l \rangle}(2)]^T$ is the position component of the target motion parameter estimate $\hat{\boldsymbol{\xi}}^{\langle l \rangle}$ obtained in the lth Monte Carlo run, and $N_{\text{MC}} = 50\,000$ is the total number of Monte Carlo runs. Similarly, the bias norm and RMSE for the velocity estimate are defined as

$$\text{Bias norm}_{\text{vel}} = \left\| \frac{1}{N_{\text{MC}}} \sum_{l=1}^{N_{\text{MC}}} \left(\hat{\boldsymbol{v}}_0^{\langle l \rangle} - \boldsymbol{v}_0 \right) \right\|, \tag{8.89a}$$

$$\text{RMSE}_{\text{vel}} = \left(\frac{1}{N_{\text{MC}}} \sum_{l=1}^{N_{\text{MC}}} \left\| \hat{\boldsymbol{v}}_0^{\langle l \rangle} - \boldsymbol{v}_0 \right\|^2 \right)^{1/2}, \tag{8.89b}$$

where $\hat{\boldsymbol{v}}_0^{\langle l \rangle} = [\hat{\boldsymbol{\xi}}^{\langle l \rangle}(3), \hat{\boldsymbol{\xi}}^{\langle l \rangle}(4)]^T$ is the velocity component of $\hat{\boldsymbol{\xi}}^{\langle l \rangle}$. The lower bound for the RMSE of position estimate is given by the square root of the sum of the first two diagonal elements of the CRLB matrix $\boldsymbol{C}_{\boldsymbol{\xi}}$, while the lower bound for the RMSE of velocity estimate is given by the square root of the sum of the last two diagonal elements of $\boldsymbol{C}_{\boldsymbol{\xi}}$. These bounds are used as benchmarks for the RMSE performance of the algorithms.

Simulation Example 1

Fig. 8.3 compares the bias norm and RMSE performance of the PLE, BCPLE, WIVE and MLE for various noise levels listed in Table 8.2. The results confirm the severe bias problem of the PLE which in turn leads to

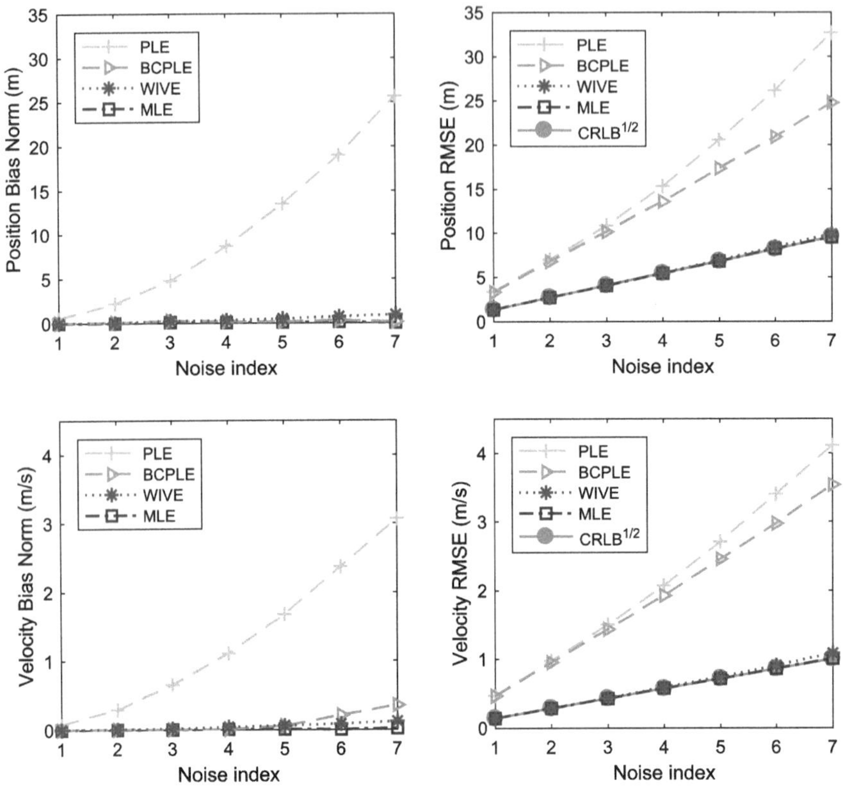

Figure 8.3 Bias norm and RMSE versus noise standard deviation in Table 8.2 for the PLE, BCPLE, WIVE and MLE in Simulation Example 1.

the poor RMSE performance of the PLE. The BCPLE exhibits a much smaller bias than the PLE thanks to the bias estimation and subtraction. However, the RMSE performance of the BCPLE is only slightly better than that of the PLE. On the other hand, the WIVE not only produces almost unbiased estimates, but also exhibits a RMSE closely matching the CRLB, thus demonstrating the performance advantages of the WIVE over the PLE and BCPLE. Note that, although showing a performance on a par with the WIVE, the MLE is computationally more expensive than the WIVE.

Simulation Example 2 (Large TDOA noise)

We now consider another simulated scenario with large TDOA noise. The measurement noise standard deviations used in this example are listed in Table 8.3. Fig. 8.4 shows the bias norm and RMSE performance of the

Table 8.3 Noise levels used in Example 2.

Noise index	1	2	3	4	5	6	7
σ_θ (degree)	1	2	3	4	5	6	7
σ_ε (m)	30	35	40	45	50	55	60
σ_ϱ (m/s)	0.1	0.2	0.3	0.4	0.5	0.6	0.7

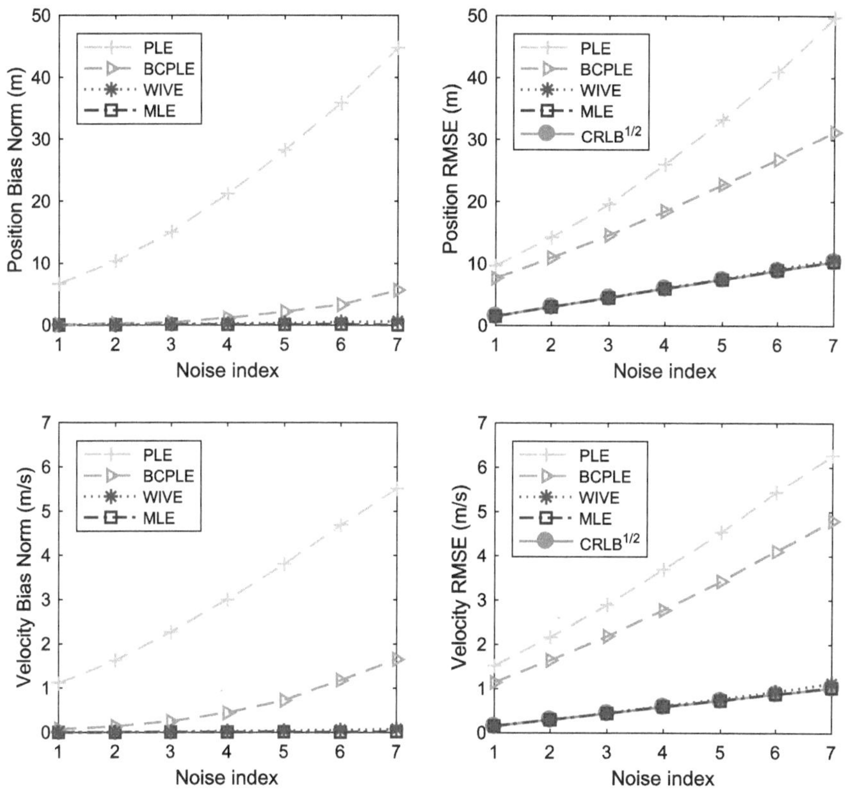

Figure 8.4 Bias norm and RMSE versus noise standard deviation in Table 8.3 for the PLE, BCPLE, WIVE and MLE in Simulation Example 2.

algorithms for this scenario. Similar to the previous example, the PLE exhibits the worst performance among the algorithms both in terms of bias and RMSE. Interestingly, in this large noise scenario, the BCPLE is observed to exhibit a notable bias which increases rapidly as the noise level increases. The reason for this is that the BCPLE relies on the PLE instantaneous bias estimation which becomes less accurate at large noise levels. In contrast, the WIVE and MLE appear to perform well even in large noise by exhibiting almost no bias while closely attaining the CRLB.

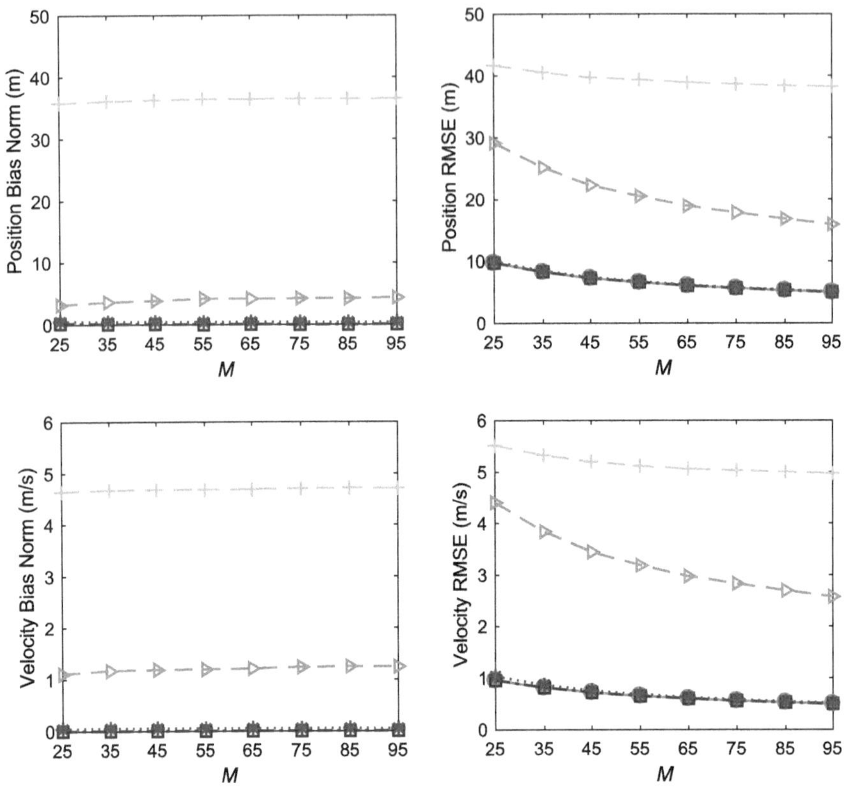

Figure 8.5 Bias norm and RMSE performance versus M for $\sigma_\theta = 6^\circ, \sigma_\varepsilon = 55$ m and $\sigma_\varrho = 0.6$ m/s (same legends as in Figs. 8.3 and 8.4 are used). Here, the observation interval length is kept unchanged ($T = 14.5$ s), while M is increased by reducing the time step $\Delta_T = t_{k+1} - t_k = T/(M-1)$ accordingly.

Fig. 8.5 plots the bias norm and RMSE versus M for noise index 6. Similar observations to Fig. 8.4 can be made here, confirming again the performance advantages of the WIVE. It is also observed in Fig. 8.5 that we obtain a better RMSE performance as M increases, as expected, because more measurements are available. However, having more measurements comes at the expense of increased computational complexity.

8.10 Summary

This chapter has considered the problem of passive multistatic TMA using AOA, TDOA and FDOA measurements. Four batch estimators were presented; viz., the PLE, BCPLE, WIVE and MLE. In contrast to the

MLE which is computationally demanding because of its iterative nature, the PLE, BCPLE and WIVE are closed-form with inherent stability and computation advantages. The PLE suffers from severe bias problems due to the correlation between the measurement matrix and pseudolinear noise vector. The BCPLE alleviates the PLE bias by estimating and subtracting the PLE instantaneous bias. The BCPLE estimate is incorporated into the WIVE to produce asymptotically unbiased estimates of target motion parameters. A theoretical covariance analysis was presented to show that the WIVE is asymptotically efficient under the assumption of small measurement noise. The performance advantages of the WIVE were corroborated by numerical simulation examples, where the WIVE was observed to produce nearly unbiased estimates with RMSE closely achieving the CRLB.

8.11 Appendices

Appendix A

This appendix shows that $\boldsymbol{F}_\circ^T \boldsymbol{W}_\circ^{-1} \boldsymbol{F}_\circ = \boldsymbol{J}_\circ^T \boldsymbol{K}^{-1} \boldsymbol{J}_\circ$. To do that, the Jacobian matrix of $\boldsymbol{e}$ with respect to $\boldsymbol{\eta}$ (denoted as $\boldsymbol{V} = \partial \boldsymbol{e}/\partial \boldsymbol{\eta}$,) is introduced:

$$\boldsymbol{V} = \text{diag}(\boldsymbol{V}_0, \boldsymbol{V}_1, \ldots, \boldsymbol{V}_{M-1}) \tag{8.90}$$

where

$$\boldsymbol{V}_k = \left[\boldsymbol{V}_{\theta,k}^T, \boldsymbol{V}_{\gamma,k}^T, \boldsymbol{V}_{\zeta,k}^T\right]^T. \tag{8.91}$$

The matrix $\boldsymbol{V}_k$ has entries given by

$$\boldsymbol{V}_{\theta,k} = \left[\mathfrak{A}_k, \boldsymbol{0}_{(N-1)\times(N-1)}, \boldsymbol{0}_{(N-1)\times(N-1)}\right], \tag{8.92a}$$

$$\boldsymbol{V}_{\gamma,k} = \left[\boldsymbol{0}_{(N-1)\times 1}, \mathfrak{B}_k, \mathfrak{C}_k, \boldsymbol{0}_{(N-1)\times(N-1)}\right], \tag{8.92b}$$

$$\boldsymbol{V}_{\zeta,k} = \left[\mathfrak{D}_k, \mathfrak{E}_k, \boldsymbol{0}_{(N-1)\times(N-1)}, -\boldsymbol{I}_{(N-1)\times(N-1)}\right], \tag{8.92c}$$

where

$$\mathfrak{A}_k = \text{diag}\left(\ldots, \|\boldsymbol{d}_{i,k}\|, \ldots\right)_{i=1,\ldots,N}, \tag{8.93a}$$

$$\mathfrak{B}_k = \text{diag}\left(\ldots, \frac{\|\boldsymbol{d}_{i,k}\|\left(\|\boldsymbol{d}_{i,k}\| - \|\boldsymbol{d}_{1,k}\|\right)}{\|\boldsymbol{r}_{i1}\|}\sin(\theta_{i,k} - \theta_{i1}), \ldots\right)_{i=2,\ldots,N}, \tag{8.93b}$$

$$\mathfrak{C}_k = \text{diag}\left(\ldots, \frac{\|\boldsymbol{d}_{1,k}\|}{\|\boldsymbol{r}_{i1}\|}\cos(\theta_{i,k} - \theta_{i1}), \ldots\right)_{i=1,\ldots,N}, \tag{8.93c}$$

$$\mathfrak{D}_k = -(v_{y,k}\cos\theta_{1,k} - v_{x,k}\sin\theta_{1,k}) \times \mathbf{1}_{(N-1)\times 1}, \tag{8.93d}$$

$$\mathfrak{E}_k = \text{diag}\left(\ldots, v_{y,k}\cos\theta_{i,k} - v_{x,k}\sin\theta_{i,k}, \ldots\right)_{i=2,\ldots,N}. \tag{8.93e}$$

It is straightforward to verify, through some algebraic manipulations, that $\boldsymbol{F}_{\circ,k} = -\boldsymbol{V}_k\boldsymbol{J}_{\circ,k}$ and $\boldsymbol{W}_k = \boldsymbol{V}_k\boldsymbol{K}_k\boldsymbol{V}_k^T$ for $k = 0, \ldots, M-1$, where $\boldsymbol{F}_{\circ,k}$ and $\boldsymbol{J}_{\circ,k}$ are the noise-free versions of $\boldsymbol{F}_k$ and $\boldsymbol{J}_k$ in (8.42) and (8.23), respectively, and are functions of the true target motion parameter vector $\boldsymbol{\xi}$. As a result, we have $\boldsymbol{F}_\circ = -\boldsymbol{V}\boldsymbol{J}_\circ$ and $\boldsymbol{W}_\circ = \boldsymbol{V}\boldsymbol{K}\boldsymbol{V}^T$, which implies

$$\begin{aligned} \boldsymbol{J}_\circ^T\boldsymbol{K}^{-1}\boldsymbol{J}_\circ &= \boldsymbol{J}_\circ^T\boldsymbol{V}^T(\boldsymbol{V}^T)^{-1}\boldsymbol{K}^{-1}\boldsymbol{V}^{-1}\boldsymbol{V}\boldsymbol{J}_\circ \\ &= (\boldsymbol{V}\boldsymbol{J}_\circ)^T(\boldsymbol{V}\boldsymbol{K}\boldsymbol{V}^T)^{-1}(\boldsymbol{V}\boldsymbol{J}_\circ) \\ &= \boldsymbol{F}_\circ^T\boldsymbol{W}_\circ^{-1}\boldsymbol{F}_\circ. \end{aligned} \tag{8.94}$$

Appendix B

From (8.37), we have

$$e_{\gamma_{i1,k}} = \boldsymbol{F}_{\gamma_{i1,k}}\boldsymbol{\xi} - b_{\gamma_{i1,k}}. \tag{8.95}$$

Substituting $\tilde{\gamma}_{i1,k} = \gamma_{i1,k} + \varepsilon_{i1,k}$ and $\tilde{\theta}_{i,k} = \theta_{i,k} + n_{i,k}$ into (8.95), and using the relationship in (8.36), we obtain an exact expression for $e_{\gamma_{i1,k}}$ as

$$\begin{aligned} e_{\gamma_{i1,k}} =& \frac{\|\boldsymbol{d}_{1,k}\|}{\|\boldsymbol{r}_{i1}\|}\cos(\theta_{i,k} - \theta_{i1})\varepsilon_{i1,k} + \frac{\|\boldsymbol{d}_{i,k}\|\left(\|\boldsymbol{d}_{i,k}\| - \|\boldsymbol{d}_{1,k}\|\right)}{\|\boldsymbol{r}_{i1}\|}\sin(\theta_{i,k} - \theta_{i1})n_{i,k} \\ &- \frac{\cos(\theta_{i,k} - \theta_{i1})}{2\|\boldsymbol{r}_{i1}\|}\varepsilon_{i1,k}^2 + \frac{2\|\boldsymbol{d}_{i,k}\|\left(\|\boldsymbol{d}_{i,k}\| - \|\boldsymbol{d}_{1,k}\|\right)}{\|\boldsymbol{r}_{i1}\|} \\ &\times \cos(\theta_{i,k} - \theta_{i1})\sin^2\left(\frac{n_{i,k}}{2}\right) \\ &+ \left(\frac{2\left(\|\boldsymbol{d}_{i,k}\| - \|\boldsymbol{d}_{1,k}\|\right)\varepsilon_{i1,k} + \varepsilon_{i1,k}^2}{\|\boldsymbol{r}_{i1}\|}\right) \\ &\times \left(\frac{1}{2}\sin(\theta_{i,k} - \theta_{i1})\sin n_{i,k} + \cos(\theta_{i,k} - \theta_{i1})\sin^2\left(\frac{n_{i,k}}{2}\right)\right). \end{aligned} \tag{8.96}$$

Similarly, from (8.40), we have

$$e_{\zeta_{i1,k}} = \boldsymbol{F}_{\zeta_{i1,k}}\boldsymbol{\xi} - b_{\zeta_{i1,k}}. \tag{8.97}$$

Substituting $\tilde{\zeta}_{i1,k} = \zeta_{i1,k} + \varrho_{i1,k}$, $\tilde{\theta}_{i,k} = \theta_{i,k} + n_{i,k}$ and $\tilde{\theta}_{1,k} = \theta_{1,k} + n_{1,k}$ into (8.97), and using relationship in (8.39), we can express $e_{\zeta_{i1,k}}$ as

$$
\begin{aligned}
e_{\zeta_{i1,k}} = & \left(v_{y,k}\cos\theta_{i,k} - v_{x,k}\sin\theta_{i,k}\right)\sin n_{i,k} \\
& - \left(v_{y,k}\cos\theta_{1,k} - v_{x,k}\sin\theta_{1,k}\right)\sin n_{1,k} - \varrho_{i1,k} \\
& - 2\left(v_{x,k}\cos\theta_{i,k} + v_{y,k}\sin\theta_{i,k}\right)\sin^2\left(\frac{n_{i,k}}{2}\right) \\
& + 2\left(v_{x,k}\cos\theta_{1,k} + v_{y,k}\sin\theta_{1,k}\right)\sin^2\left(\frac{n_{1,k}}{2}\right).
\end{aligned}
\tag{8.98}
$$

Appendix C

From (8.23) and (8.26), the CRLB can be expressed as

$$
\boldsymbol{C}_{\boldsymbol{\xi}}^{-1} = \boldsymbol{J}_{\circ}^{T}\boldsymbol{K}^{-1}\boldsymbol{J}_{\circ} = \sum_{k=0}^{M-1}\boldsymbol{J}_{\circ,k}^{T}\boldsymbol{K}_{k}^{-1}\boldsymbol{J}_{\circ,k}
\tag{8.99}
$$

where $\boldsymbol{J}_{\circ,k}$ is the noise-free version of $\boldsymbol{J}_k$ in (8.23) as a function of the true value of $\boldsymbol{\xi}$. We can express $\boldsymbol{J}_{\circ,k}$ and $\boldsymbol{K}_k$ explicitly in terms of AOA and TDOA/FDOA components as

$$
\boldsymbol{J}_{\circ,k} = \begin{bmatrix} \boldsymbol{J}_{\circ,k}^{\mathrm{AOA}} \\ \boldsymbol{J}_{\circ,k}^{\mathrm{TDOA/FDOA}} \end{bmatrix}, \boldsymbol{K}_k = \begin{bmatrix} \boldsymbol{K}_k^{\mathrm{AOA}} & \boldsymbol{0} \\ \boldsymbol{0} & \boldsymbol{K}_k^{\mathrm{TDOA/FDOA}} \end{bmatrix}.
$$

Therefore, (8.99) becomes

$$
\begin{aligned}
\boldsymbol{C}_{\boldsymbol{\xi}}^{-1} = & \sum_{k=0}^{M-1}\left(\boldsymbol{J}_{\circ,k}^{\mathrm{AOA}}\right)^{T}\left(\boldsymbol{K}_{k}^{\mathrm{AOA}}\right)^{-1}\boldsymbol{J}_{\circ,k}^{\mathrm{AOA}} \\
& + \sum_{k=0}^{M-1}\left(\boldsymbol{J}_{\circ,k}^{\mathrm{TDOA/FDOA}}\right)^{T}\left(\boldsymbol{K}_{k}^{\mathrm{TDOA/FDOA}}\right)^{-1}\boldsymbol{J}_{\circ,k}^{\mathrm{TDOA/FDOA}}.
\end{aligned}
\tag{8.100}
$$

Since the choice of TDOA/FDOA reference receiver does not affect the second term on the right-hand side of (8.100) [86], the CRLB matrix $\boldsymbol{C}_{\boldsymbol{\xi}}$ is independent of the choice of TDOA/FDOA reference receiver.

CHAPTER 9

Closed-form solutions for multistatic target localization with time-difference-of-arrival measurements

Contents

9.1 Introduction

In this chapter, we consider the problem of target localization using TDOA measurements collected by a passive multistatic radar system. The radar-target geometry is depicted in Fig. 9.1. The radar system consists of N receivers located at $\boldsymbol{r}_i$, $i=1,2,\ldots,N$. The objective of the target localization problem under consideration is to estimate the target position $\boldsymbol{p}$ using TDOA measurements of radar signals arriving at the receivers. Note that $\boldsymbol{r}_i$ and $\boldsymbol{p}$ are $D \times 1$ column vectors of Cartesian coordinates, where $D=2$ or 3 is the space dimension.

The TDOA measurement between receiver i and receiver 1 is given by

$$\tilde{d}_{i1} = d_{i1} + n_{i1}, \quad i=2,3,\ldots,N, \tag{9.1}$$

where d_{i1} is the true TDOA and n_{i1} is the measurement noise. The true TDOA d_{i1} is related to the target position $\boldsymbol{p}$ and receiver positions $\boldsymbol{r}_i$ and $\boldsymbol{r}_1$ by

$$d_{i1} = d_i - d_1 = \|\boldsymbol{p} - \boldsymbol{r}_i\| - \|\boldsymbol{p} - \boldsymbol{r}_1\|. \tag{9.2}$$

Signal Processing for Multistatic Radar Systems
https://doi.org/10.1016/B978-0-12-815314-7.00019-6

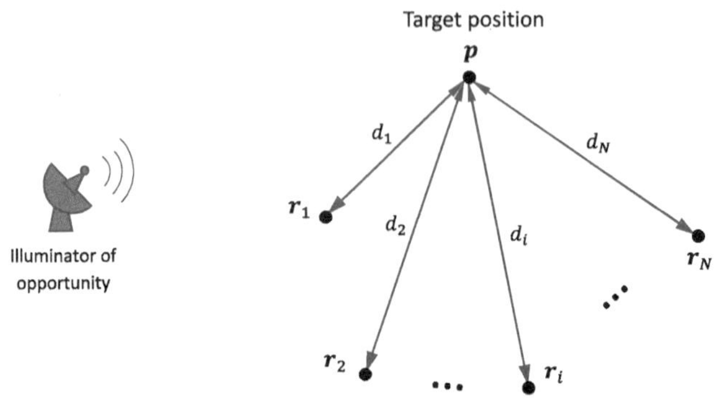

Figure 9.1 Target localization geometry with a passive multistatic radar system using TDOA measurements.

Here, we have already normalized the TDOA equation by the signal propagation speed c. Although the TDOA measurements can be obtained via any non-redundant set of $N-1$ receiver pairs, it is common practice to use a fixed receiver as the reference receiver (e.g., receiver 1 in this case).

Stacking $N-1$ TDOA measurements for $i=2,3,\ldots,N$ yields

$$\tilde{\boldsymbol{d}} = \boldsymbol{d} + \boldsymbol{n} \tag{9.3}$$

where

$$\tilde{\boldsymbol{d}} = [\tilde{d}_{21}, \tilde{d}_{31}, \ldots, \tilde{d}_{N1}]^T, \tag{9.4a}$$

$$\boldsymbol{d} = [d_{21}, d_{31}, \ldots, d_{N1}]^T, \tag{9.4b}$$

$$\boldsymbol{n} = [n_{21}, n_{31}, \ldots, n_{N1}]^T. \tag{9.4c}$$

The noise vector $\boldsymbol{n}$ is modeled as a zero-mean Gaussian with covariance matrix $\mathbf{Q}$. In this chapter, we assume, as is often the case in practice, that the exact expression for $\mathbf{Q}$ is not known but its structure is available *a priori*. For example, $\mathbf{Q}$ is proportional to $(\boldsymbol{I}+\mathbf{1})$ for the case of uncorrelated noise with equal SNR across the receivers [108].

The TDOA localization problem can be intuitively viewed as finding the intersection of the hyperbolic surfaces defined by TDOA measurements. Numerous techniques have been proposed in the literature for TDOA localization. These can be broadly categorized into three main approaches:

- maximum-likelihood estimation [74]
- intersection of hyperbolic asymptotes [109–111]

- linearized least-squares with parameter constraint [69,108,112–115].

The first approach relies on the maximum-likelihood principle which is a popular technique for solving nonlinear estimation problems. Although the MLE is asymptotically unbiased and efficient, it does not admit a closed-form solution and its solution must be obtained via numerical search algorithms which are computationally expensive. Moreover, the MLE is prone to convergence to local minima or divergence because of the nonconvexity of the maximum-likelihood cost function. In addition, being a nonlinear estimator it is vulnerable to the threshold effect.

The second approach exploits the intersection of the asymptotes of TDOA hyperbolae which allows closed-form solutions to be developed. Specifically, based on the use of hyperbolic asymptotes, the TDOA localization problem is reformulated as a new AOA localization problem which can efficiently be solved via closed-form pseudolinear techniques. However, this approach was originally developed for TDOA localization in 2D and its extension to 3D is not straightforward.

The third approach, using linearized least-squares with parameter constraint, which also provides closed-form solutions and more importantly is applicable to both 2D and 3D cases, is an attractive alternative to the aforementioned approaches and will be the focus of this chapter. The main idea of this approach is to algebraically rearrange the nonlinear TDOA measurement equations into a set of linear equations with respect to the target position by introducing a nuisance parameter. This then enables the use of linear least-squares for target position estimation. The three-stage solution developed in [108] is one of the most popular algorithms of this kind. This solution also takes into account the dependence between the nuisance parameter and target position in order to enhance the accuracy of the linear least-squares solution. Unfortunately, the transformation of nonlinear TDOA measurement equations to a set of equations linear in the unknowns causes the measurement matrix in the linearized equations to be correlated with the pseudolinear noise vector. Consequently, this creates an estimation bias in the three-stage solution [114]. To overcome this bias issue, several methods have been reported. The work in [112] incorporated a constraint for the unknowns directly into the least-squares minimization, but a numerical search is required. The method of total least-squares (TLS) was exploited in [115] to tackle the correlation between the measurement matrix and pseudolinear noise vector. However, this technique yields a larger error covariance than the three-stage solution [108]. Recently, two bias reduction methods have been developed: (i) the augmented solution with quadratic constraint [114] and (ii) the instrumental variable-based solution [116]. These two methods are both capable of achieving a significant bias reduction while at the same time maintaining the optimum performance in terms of mean-squared error by closely attaining the CRLB. The instrumental variable-based solution has an advantage over the augmented solution in that it is computationally simpler than the augmented solution.

9.2 Maximum likelihood estimator and Cramér–Rao lower bound

The likelihood function for the measurement vector $\tilde{\boldsymbol{d}}$ under the Gaussian noise assumption is a multivariate Gaussian probability density function given by

$$\mathcal{L}(\tilde{\boldsymbol{d}}|\boldsymbol{p}) = \frac{1}{(2\pi)^{\frac{(N-1)}{2}}|\mathbf{Q}|^{\frac{1}{2}}} \exp\left\{-\frac{1}{2}\left(\tilde{\boldsymbol{d}} - \boldsymbol{d}(\boldsymbol{p})\right)^T \mathbf{Q}^{-1}\left(\tilde{\boldsymbol{d}} - \boldsymbol{d}(\boldsymbol{p})\right)\right\}. \quad (9.5)$$

Note that $\boldsymbol{d}(\boldsymbol{p})$ is written explicitly as a function of $\boldsymbol{p}$.

The MLE aims to maximize the log-likelihood function $\ln \mathcal{L}(\tilde{\boldsymbol{d}}|\boldsymbol{p})$ over $\boldsymbol{p}$, which is equivalent to

$$\hat{\boldsymbol{p}}_{\mathrm{MLE}} = \arg\min_{\boldsymbol{p}\in\mathbb{R}^D} J_{\mathrm{ML}}(\boldsymbol{p}) \quad (9.6)$$

where the objective function $J_{\mathrm{ML}}(\boldsymbol{p})$ is given by

$$J_{\mathrm{ML}}(\boldsymbol{p}) = \frac{1}{2}\left(\tilde{\boldsymbol{d}} - \boldsymbol{d}(\boldsymbol{p})\right)^T \mathbf{Q}^{-1}\left(\tilde{\boldsymbol{d}} - \boldsymbol{d}(\boldsymbol{p})\right). \quad (9.7)$$

This minimization is a nonlinear least-squares estimation problem which does not admit a closed-form solution. Numerical solutions for (9.6) can be obtained via various gradient or downhill simplex-based iterative search algorithms such as the Gauss–Newton algorithm [93], the steepest-descent algorithm [94], and the Nelder–Mead simplex algorithm [95]. The Gauss–Newton implementation of the MLE is given by the iterations

$$\hat{\boldsymbol{p}}(j+1) = \hat{\boldsymbol{p}}(j) + \left(\boldsymbol{J}^T(j)\mathbf{Q}^{-1}\boldsymbol{J}(j)\right)^{-1}\boldsymbol{J}^T(j)\mathbf{Q}^{-1}\left(\tilde{\boldsymbol{d}} - \boldsymbol{d}(\hat{\boldsymbol{p}})\right), \quad j = 0, 1, \ldots, \quad (9.8)$$

where $\boldsymbol{J}(j)$ is the Jacobian matrix $\boldsymbol{J}(\boldsymbol{p})$ of $\boldsymbol{d}(\boldsymbol{p})$ with respect to $\boldsymbol{p}$ evaluated at $\boldsymbol{p} = \hat{\boldsymbol{p}}(j)$. The Jacobian matrix as a function of $\boldsymbol{p}$ is given by

$$\boldsymbol{J}(\boldsymbol{p}) = [\boldsymbol{J}_{21}^T(\boldsymbol{p}), \boldsymbol{J}_{31}^T(\boldsymbol{p}), \ldots, \boldsymbol{J}_{N1}^T(\boldsymbol{p})]^T \tag{9.9}$$

where

$$\boldsymbol{J}_{i1}(\boldsymbol{p}) = \left(\frac{\boldsymbol{p} - \boldsymbol{r}_i}{\|\boldsymbol{p} - \boldsymbol{r}_i\|} - \frac{\boldsymbol{p} - \boldsymbol{r}_1}{\|\boldsymbol{p} - \boldsymbol{r}_1\|} \right)^T. \tag{9.10}$$

Despite the asymptotic unbiasedness and efficiency of the MLE, the iterative solutions for the MLE are not only computationally demanding, but also susceptible to divergence due to the non-convex nature of the maximum-likelihood cost function.

The CRLB for the TDOA localization problem under consideration is given by

$$\boldsymbol{C}_{\mathrm{CRLB}} = \left(\boldsymbol{J}^T(\boldsymbol{p}) \boldsymbol{Q}^{-1} \boldsymbol{J}(\boldsymbol{p})\right)^{-1} \tag{9.11}$$

where the Jacobian matrix $\boldsymbol{J}(\boldsymbol{p})$ is computed from the true target position $\boldsymbol{p}$.

9.3 Three-stage least-squares solution

The closed-form solution [108] consists of three stages. In the first stage, the TDOA equations are algebraically transformed into a system of linear equations based on an introduction of a nuisance parameter. A linear least-squares solution for the unknown vector of target position and nuisance parameter is then obtained. The second stage exploits the constraint relationship between the target position and nuisance parameter to enhance estimation accuracy. In the third stage, the solution of the second stage is mapped onto the final target position estimate.

Stage 1

From (9.2), we have

$$(\|\boldsymbol{p} - \boldsymbol{r}_i\|)^2 = (d_{i1} + \|\boldsymbol{p} - \boldsymbol{r}_1\|)^2, \tag{9.12}$$

which, after some algebraic manipulations, becomes

$$-2(\boldsymbol{r}_i - \boldsymbol{r}_1)^T \boldsymbol{p} - 2d_{i1}\|\boldsymbol{p} - \boldsymbol{r}_1\| = d_{i1}^2 - \boldsymbol{r}_i^T \boldsymbol{r}_i + \boldsymbol{r}_1^T \boldsymbol{r}_1. \tag{9.13}$$

Replacing the true TDOA d_{i1} with its noisy measurement $\tilde{d}_{i1}$ in (9.13) yields

$$-2(\boldsymbol{r}_i-\boldsymbol{r}_1)^T\boldsymbol{p}-2\tilde{d}_{i1}\|\boldsymbol{p}-\boldsymbol{r}_1\| = \tilde{d}_{i1}^2-\boldsymbol{r}_i^T\boldsymbol{r}_i+\boldsymbol{r}_1^T\boldsymbol{r}_1-\left(2\|\boldsymbol{p}-\boldsymbol{r}_i\|n_{i1}+n_{i1}^2\right). \quad (9.14)$$

Stacking this equation for $i = 2, 3, \ldots, N$, we obtain

$$\mathbf{G}_1\boldsymbol{\varphi}_1 = \boldsymbol{h}_1 + \boldsymbol{\eta}_1 \quad (9.15)$$

where

$$\boldsymbol{\varphi}_1 = [\boldsymbol{p}^T, d_1]^T,\ d_1 = \|\boldsymbol{p}-\boldsymbol{r}_1\|, \quad (9.16)$$

$$\mathbf{G}_1 = -2\begin{bmatrix} (\boldsymbol{r}_2-\boldsymbol{r}_1)^T & \tilde{d}_{21} \\ (\boldsymbol{r}_3-\boldsymbol{r}_1)^T & \tilde{d}_{31} \\ \vdots & \vdots \\ (\boldsymbol{r}_N-\boldsymbol{r}_1)^T & \tilde{d}_{N1} \end{bmatrix}, \quad (9.17)$$

$$\boldsymbol{h}_1 = \begin{bmatrix} \tilde{d}_{21}^2 - \boldsymbol{r}_2^T\boldsymbol{r}_2 + \boldsymbol{r}_1^T\boldsymbol{r}_1 \\ \tilde{d}_{31}^2 - \boldsymbol{r}_3^T\boldsymbol{r}_3 + \boldsymbol{r}_1^T\boldsymbol{r}_1 \\ \vdots \\ \tilde{d}_{N1}^2 - \boldsymbol{r}_N^T\boldsymbol{r}_N + \boldsymbol{r}_1^T\boldsymbol{r}_1 \end{bmatrix}, \quad (9.18)$$

$$\boldsymbol{\eta}_1 = -\begin{bmatrix} 2\|\boldsymbol{p}-\boldsymbol{r}_2\|n_{21} + n_{21}^2 \\ 2\|\boldsymbol{p}-\boldsymbol{r}_3\|n_{31} + n_{31}^2 \\ \vdots \\ 2\|\boldsymbol{p}-\boldsymbol{r}_N\|n_{N1} + n_{N1}^2 \end{bmatrix}. \quad (9.19)$$

Eq. (9.15) can be referred to as the pseudolinear equations, where $\boldsymbol{\varphi}_1$ is a unknown vector consisting of the target position $\boldsymbol{p}$ and a nuisance parameter d_1 which is also dependent on $\boldsymbol{p}$. Here, the matrix $\mathbf{G}_1$ can be referred to as the measurement matrix and the vector $\boldsymbol{\eta}_1$ as the pseudolinear noise vector. Ignoring the second-order noise terms in (9.19), the covariance matrix of $\boldsymbol{\eta}_1$ can be approximated by

$$\boldsymbol{C}_{\eta_1} = E\left\{\boldsymbol{\eta}_1\boldsymbol{\eta}_1^T\right\} \approx \boldsymbol{B}_1^{\circ}\mathbf{Q}\boldsymbol{B}_1^{\circ} \quad (9.20)$$

where

$$\boldsymbol{B}_1^{\circ} = 2\,\mathrm{diag}(\|\boldsymbol{p}-\boldsymbol{r}_2\|, \|\boldsymbol{p}-\boldsymbol{r}_3\|, \ldots, \|\boldsymbol{p}-\boldsymbol{r}_N\|). \quad (9.21)$$

We can first solve (9.15) for $\boldsymbol{\varphi}_1$ and extract the first D elements of $\boldsymbol{\varphi}_1$ to obtain an estimate of $\boldsymbol{p}$. Solving (9.15) in the least-squares sense gives

$$\hat{\boldsymbol{\varphi}}_1 = (\boldsymbol{G}_1^T \boldsymbol{W}_1 \boldsymbol{G}_1)^{-1} \boldsymbol{G}_1^T \boldsymbol{W}_1 \boldsymbol{h}_1. \tag{9.22}$$

Ideally, the weighting matrix $\boldsymbol{W}_1$ is given by the inverse of the noise covariance $\boldsymbol{C}_{\boldsymbol{\eta}_1}$. However, since $\boldsymbol{B}_1^\circ$ is a function of the unknown target position $\boldsymbol{p}$, $\boldsymbol{C}_{\boldsymbol{\eta}_1}$ cannot be computed. Therefore, an initial estimate of $\boldsymbol{\varphi}_1$, denoted as $\hat{\boldsymbol{\varphi}}_1^{\text{init}}$, is computed from (9.22) by setting $\boldsymbol{W}_1 = \boldsymbol{Q}^{-1}$ and then used to approximate $\boldsymbol{B}_1^\circ$ as

$$\boldsymbol{B}_1 = 2\,\text{diag}\left(\|\hat{\boldsymbol{\varphi}}_1^{\text{init}}(1:D) - \boldsymbol{r}_2\|, \|\hat{\boldsymbol{\varphi}}_1^{\text{init}}(1:D) - \boldsymbol{r}_3\|, \ldots, \|\hat{\boldsymbol{\varphi}}_1^{\text{init}}(1:D) - \boldsymbol{r}_N\|\right). \tag{9.23}$$

This matrix is then used to compute the weighting matrix $\boldsymbol{W}_1$ in (9.22):

$$\boldsymbol{W}_1 = (\boldsymbol{B}_1 \boldsymbol{Q} \boldsymbol{B}_1)^{-1}. \tag{9.24}$$

The estimation error of $\hat{\boldsymbol{\varphi}}_1$ is given by

$$\begin{aligned}
\Delta\boldsymbol{\varphi}_1 &= \hat{\boldsymbol{\varphi}}_1 - \boldsymbol{\varphi}_1 && (9.25a)\\
&= (\boldsymbol{G}_1^T \boldsymbol{W}_1 \boldsymbol{G}_1)^{-1} \boldsymbol{G}_1^T \boldsymbol{W}_1 \left(\boldsymbol{h}_1 - \boldsymbol{G}_1 \boldsymbol{\varphi}_1\right) && (9.25b)\\
&= -(\boldsymbol{G}_1^T \boldsymbol{W}_1 \boldsymbol{G}_1)^{-1} \boldsymbol{G}_1^T \boldsymbol{W}_1 \boldsymbol{\eta}_1 && (9.25c)\\
&= (\boldsymbol{G}_1^T \boldsymbol{W}_1 \boldsymbol{G}_1)^{-1} \boldsymbol{G}_1^T \boldsymbol{W}_1 \left(\boldsymbol{B}_1^\circ \boldsymbol{n} + \boldsymbol{n} \odot \boldsymbol{n}\right) && (9.25d)
\end{aligned}$$

and the covariance matrix of $\Delta\boldsymbol{\varphi}_1$ can be approximated by

$$\boldsymbol{C}_{\boldsymbol{\varphi}_1} = E\left\{\Delta\boldsymbol{\varphi}_1 \Delta\boldsymbol{\varphi}_1^T\right\} \approx \left(\boldsymbol{G}_1^{\circ T} \boldsymbol{W}_1^\circ \boldsymbol{G}_1^\circ\right)^{-1}. \tag{9.26}$$

Here, $\boldsymbol{G}_1^\circ$ and $\boldsymbol{W}_1^\circ$ are the noise-free versions of $\boldsymbol{G}_1$ and $\boldsymbol{W}_1$, respectively.

Stage 2

The solution $\hat{\boldsymbol{\varphi}}_1$ in (9.22) assumes that $\boldsymbol{p}$ and d_1 are independent. However, this is not the case because d_1 is a function of $\boldsymbol{p}$. In order to incorporate the relationship between d_1 and $\boldsymbol{p}$ into the estimation, we can formulate another set of linear equations by subtracting $\boldsymbol{r}_1$ from the first D elements of $\boldsymbol{\varphi}_1$ and taking square the elements, thus leading to

$$\begin{bmatrix} \boldsymbol{I}_{D\times D} \\ \boldsymbol{1}_{1\times D} \end{bmatrix} \left((\boldsymbol{p} - \boldsymbol{r}_1) \odot (\boldsymbol{p} - \boldsymbol{r}_1)\right) = \left(\boldsymbol{\varphi}_1 - \begin{bmatrix} \boldsymbol{r}_1 \\ 0 \end{bmatrix}\right) \odot \left(\boldsymbol{\varphi}_1 - \begin{bmatrix} \boldsymbol{r}_1 \\ 0 \end{bmatrix}\right) \tag{9.27}$$

where $\odot$ stands for the element-by-element (Schur) product. Replacing $\boldsymbol{\varphi}_1$ with its estimate $\hat{\boldsymbol{\varphi}}_1$ given by (9.22), we obtain

$$\begin{bmatrix} \boldsymbol{I}_{D\times D} \\ \boldsymbol{1}_{1\times D} \end{bmatrix} ((\boldsymbol{p}-\boldsymbol{r}_1)\odot(\boldsymbol{p}-\boldsymbol{r}_1)) = \left(\hat{\boldsymbol{\varphi}}_1 - \begin{bmatrix} \boldsymbol{r}_1 \\ 0 \end{bmatrix}\right) \odot \left(\hat{\boldsymbol{\varphi}}_1 - \begin{bmatrix} \boldsymbol{r}_1 \\ 0 \end{bmatrix}\right) - 2\,\mathrm{diag}\left(\boldsymbol{\varphi}_1 - \begin{bmatrix} \boldsymbol{r}_1 \\ 0 \end{bmatrix}\right)\Delta\boldsymbol{\varphi}_1 - \Delta\boldsymbol{\varphi}_1 \odot \Delta\boldsymbol{\varphi}_1, \tag{9.28}$$

which can be succinctly written as

$$\boldsymbol{G}_2\boldsymbol{\varphi}_2 = \boldsymbol{h}_2 + \boldsymbol{\eta}_2 \tag{9.29}$$

with

$$\boldsymbol{\varphi}_2 = (\boldsymbol{p}-\boldsymbol{r}_1)\odot(\boldsymbol{p}-\boldsymbol{r}_1), \tag{9.30}$$

$$\boldsymbol{G}_2 = \begin{bmatrix} \boldsymbol{I}_{D\times D} \\ \boldsymbol{1}_{1\times D} \end{bmatrix}, \tag{9.31}$$

$$\boldsymbol{h}_2 = \left(\hat{\boldsymbol{\varphi}}_1 - \begin{bmatrix} \boldsymbol{r}_1 \\ 0 \end{bmatrix}\right) \odot \left(\hat{\boldsymbol{\varphi}}_1 - \begin{bmatrix} \boldsymbol{r}_1 \\ 0 \end{bmatrix}\right), \tag{9.32}$$

$$\boldsymbol{\eta}_2 = -2\,\mathrm{diag}\left(\boldsymbol{\varphi}_1 - \begin{bmatrix} \boldsymbol{r}_1 \\ 0 \end{bmatrix}\right)\Delta\boldsymbol{\varphi}_1 - \Delta\boldsymbol{\varphi}_1 \odot \Delta\boldsymbol{\varphi}_1. \tag{9.33}$$

Here, $\boldsymbol{\eta}_2$ is the pseudolinear noise term. Ignoring the second-order noise terms in (9.33), the covariance matrix of $\boldsymbol{\eta}_2$ can be approximated as

$$\boldsymbol{C}_{\eta_2} = E\{\boldsymbol{\eta}_2\boldsymbol{\eta}_2^T\} \approx \boldsymbol{B}_2^\circ \boldsymbol{C}_{\varphi_1} \boldsymbol{B}_2^\circ \tag{9.34}$$

where

$$\boldsymbol{B}_2^\circ = 2\,\mathrm{diag}\left(\boldsymbol{\varphi}_1 - \begin{bmatrix} \boldsymbol{r}_1 \\ 0 \end{bmatrix}\right). \tag{9.35}$$

The least-squares solution for (9.29) is given by

$$\hat{\boldsymbol{\varphi}}_2 = (\boldsymbol{G}_2^T\boldsymbol{W}_2\boldsymbol{G}_2)^{-1}\boldsymbol{G}_2^T\boldsymbol{W}_2\boldsymbol{h}_2 \tag{9.36}$$

where the weighting matrix is $\boldsymbol{W}_2 = \boldsymbol{C}_{\eta_2}^{-1}$. However, $\boldsymbol{C}_{\eta_2}$ is not available since it is dependent on $\boldsymbol{B}_2^\circ$ and $\boldsymbol{C}_{\varphi_1}$ which are both functions of the unknown $\boldsymbol{p}$. Therefore, an estimate of $\boldsymbol{B}_2^\circ$, denoted as $\boldsymbol{B}_2$, is calculated from $\hat{\boldsymbol{\varphi}}_1$:

$$\boldsymbol{B}_2 = 2\,\mathrm{diag}\left(\hat{\boldsymbol{\varphi}}_1 - \begin{bmatrix} \boldsymbol{r}_1 \\ 0 \end{bmatrix}\right), \tag{9.37}$$

and an estimate of $\boldsymbol{C}_{\varphi_1}$ obtained via $\mathbf{G}_1$ is utilized to calculate $\boldsymbol{C}_{\eta_2}$ and $\boldsymbol{W}_2$. As a result, we obtain

$$\boldsymbol{W}_2 = \boldsymbol{B}_2^{-1}(\mathbf{G}_1^T \boldsymbol{W}_1 \mathbf{G}_1)\boldsymbol{B}_2^{-1}. \tag{9.38}$$

The estimation error of $\hat{\boldsymbol{\varphi}}_2$ is given by

$$\begin{aligned}
\Delta\boldsymbol{\varphi}_2 &= \hat{\boldsymbol{\varphi}}_2 - \boldsymbol{\varphi}_2 && (9.39\text{a})\\
&= (\mathbf{G}_2^T \boldsymbol{W}_2 \mathbf{G}_2)^{-1}\mathbf{G}_2^T \boldsymbol{W}_2 \left(\boldsymbol{h}_2 - \mathbf{G}_2\boldsymbol{\varphi}_2\right) && (9.39\text{b})\\
&= -(\mathbf{G}_2^T \boldsymbol{W}_2 \mathbf{G}_2)^{-1}\mathbf{G}_2^T \boldsymbol{W}_2 \boldsymbol{\eta}_2 && (9.39\text{c})\\
&= (\mathbf{G}_2^T \boldsymbol{W}_2 \mathbf{G}_2)^{-1}\mathbf{G}_2^T \boldsymbol{W}_2 \left(\boldsymbol{B}_2^\circ \Delta\boldsymbol{\varphi}_1 + \Delta\boldsymbol{\varphi}_1 \odot \Delta\boldsymbol{\varphi}_1\right). && (9.39\text{d})
\end{aligned}$$

Noting that $\mathbf{G}_2$ is deterministic (i.e., not containing any noise terms), the covariance matrix of $\Delta\boldsymbol{\varphi}_2$ is given by

$$\boldsymbol{C}_{\varphi_2} = E\left\{\Delta\boldsymbol{\varphi}_2 \Delta\boldsymbol{\varphi}_2^T\right\} \approx \left(\mathbf{G}_2^T \boldsymbol{W}_2^\circ \mathbf{G}_2\right)^{-1} \tag{9.40}$$

where $\boldsymbol{W}_2^\circ$ is the noise-free version of $\boldsymbol{W}_2$.

Stage 3

Using $\hat{\boldsymbol{\varphi}}_2$, the final target position estimate is obtained from

$$\hat{\boldsymbol{p}} = \boldsymbol{\Pi}\sqrt{\hat{\boldsymbol{\varphi}}_2} + \boldsymbol{r}_1 \tag{9.41}$$

where $\boldsymbol{\Pi} = \mathrm{diag}\{\mathrm{sgn}(\hat{\boldsymbol{\varphi}}_1(1:D) - \boldsymbol{r}_1)\}$ and $\mathrm{sgn}(\cdot)$ stands for the signum function.

The estimation error of $\hat{\boldsymbol{p}}$ is defined as

$$\Delta\boldsymbol{p} = \hat{\boldsymbol{p}} - \boldsymbol{p}. \tag{9.42}$$

To derive the covariance matrix for $\Delta\boldsymbol{p}$, we note that

$$\begin{aligned}
\Delta\boldsymbol{\varphi}_2 &= \hat{\boldsymbol{\varphi}}_2 - \boldsymbol{\varphi}_2 && (9.43\text{a})\\
&= (\hat{\boldsymbol{p}} - \boldsymbol{r}_1) \odot (\hat{\boldsymbol{p}} - \boldsymbol{r}_1) - (\boldsymbol{p} - \boldsymbol{r}_1) \odot (\boldsymbol{p} - \boldsymbol{r}_1) && (9.43\text{b})\\
&= 2\,\mathrm{diag}\left(\boldsymbol{p} - \boldsymbol{r}_1\right)\Delta\boldsymbol{p} + \Delta\boldsymbol{p} \odot \Delta\boldsymbol{p}. && (9.43\text{c})
\end{aligned}$$

Ignoring the second-order noise term in (9.43), we have

$$C_p = E\{\Delta p \Delta p^T\} \approx B_3^{\circ -1} C_{\varphi_2} B_3^{\circ -1} \tag{9.44}$$

where

$$B_3^{\circ} = 2\,\mathrm{diag}\,(p - r_1)\,. \tag{9.45}$$

The error covariance matrix C_p in (9.44) was shown in [108] to be equivalent to the CRLB matrix C_{CRLB} in (9.11). This implies that, under small noise conditions, the target position estimate $\hat{p}$ in (9.41) approximately achieves the CRLB and is therefore efficient.

9.4 Bias analysis

The three-stage least-squares solution covered in the previous section suffers from a bias problem because the measurement matrix G_1 is calculated from the noisy TDOA measurements and thus is correlated with the pseudolinear noise vector η_1. Specifically, the correlation between G_1 and η_1 leads to a bias in $\hat{\varphi}_1$ which subsequently propagates to $\hat{\varphi}_2$ and $\hat{p}$.

Taking the expectation of $\Delta\varphi_1$ in (9.25) reveals the bias of $\hat{\varphi}_1$ [114]

$$\begin{aligned} \delta\varphi_1 = E\{\Delta\varphi_1\} \approx H_1\left(q + 2QB_1^{\circ}H_1(D+1,:)^T\right) \\ + 2(G_1^{\circ T} W_1^{\circ} G_1^{\circ})^{-1} \begin{bmatrix} 0_{D\times 1} \\ \mathrm{trace}\left(W_1^{\circ}(G_1^{\circ}H_1 - I)B_1^{\circ}Q\right) \end{bmatrix} \end{aligned} \tag{9.46}$$

where

$$H_1 = (G_1^{\circ T} W_1^{\circ} G_1^{\circ})^{-1} G_1^{\circ T} W_1^{\circ}. \tag{9.47}$$

Here, $H_1(D+1,:)$ is the last row of H_1 and q is a column vector containing the diagonal elements of Q. It should be noted that the first bias term $H_1 q$ in (9.46) results from the second-order noise term in η_1 while the remaining bias terms are caused by the correlation between G_1 and η_1.

The bias of $\hat{\varphi}_2$ is obtained by taking the expectation of (9.39), leading to [114]

$$\delta\varphi_2 = E\{\Delta\varphi_2\} \approx H_2\left(c_1 + B_2\delta\varphi_1 + W_2^{\circ -1}(\alpha + \beta + \gamma)\right) \tag{9.48}$$

where

$$H_2 = (G_2^T W_2^{\circ} G_2)^{-1} G_2^T W_2^{\circ}. \tag{9.49}$$

Note that the bias expression in (9.48) has taken into account the error of calculating $\boldsymbol{W}_2$ from $\hat{\boldsymbol{\varphi}}_1$. Here, $\boldsymbol{c}_1$ is a column vector which contains the diagonal elements of $\boldsymbol{C}_{\varphi_1}$. The expressions for $\boldsymbol{\alpha}$, $\boldsymbol{\beta}$ and $\boldsymbol{\gamma}$ are given by [114]

$$\boldsymbol{\alpha} = -2\boldsymbol{B}_2^{\circ-1}\left(\boldsymbol{G}_1^{\circ T}\boldsymbol{W}_1^{\circ}\boldsymbol{Q}\boldsymbol{V}_a(D+1,:)^T + \begin{bmatrix} \boldsymbol{0}_{D\times 1} \\ \mathrm{trace}\left(\boldsymbol{W}_1^{\circ}\boldsymbol{G}_1^{\circ}\boldsymbol{V}_a\boldsymbol{Q}\right)\end{bmatrix}\right), \tag{9.50a}$$

$$\boldsymbol{\beta} = -2\boldsymbol{B}_2^{\circ-1}\boldsymbol{v}_b, \tag{9.50b}$$

$$\boldsymbol{\gamma} = -2\boldsymbol{W}_2^{\circ}\boldsymbol{B}_2^{\circ-1}\boldsymbol{v}_c, \tag{9.50c}$$

where

$$\boldsymbol{V}_a = \boldsymbol{B}_2^{\circ-1}(\boldsymbol{I} - \boldsymbol{G}_2\boldsymbol{H}_2)\boldsymbol{B}_2^{\circ}\boldsymbol{H}_1\boldsymbol{B}_1^{\circ} \tag{9.51}$$

and $\boldsymbol{v}_b$ and $\boldsymbol{v}_c$ are column vectors consisting of the diagonal elements of $\boldsymbol{V}_b$ and $\boldsymbol{V}_c$, respectively, whose expressions are given by

$$\boldsymbol{V}_b = \boldsymbol{W}_2^{\circ}(\boldsymbol{I} - \boldsymbol{G}_2\boldsymbol{H}_2)\boldsymbol{B}_2^{\circ}\boldsymbol{C}_{\varphi_1}, \tag{9.52a}$$

$$\boldsymbol{V}_c = (\boldsymbol{I} - \boldsymbol{G}_2\boldsymbol{H}_2)\boldsymbol{B}_2^{\circ}\boldsymbol{C}_{\varphi_1}. \tag{9.52b}$$

To obtain the bias of the final target position estimate $\hat{\boldsymbol{p}}$, we recall from (9.43) that

$$\Delta\boldsymbol{p} = \boldsymbol{B}_3^{\circ-1}(\Delta\boldsymbol{\varphi}_2 - \Delta\boldsymbol{p}\odot\Delta\boldsymbol{p}). \tag{9.53}$$

By taking the expectation of $\Delta\boldsymbol{p}$, the bias of $\hat{\boldsymbol{p}}$ is given by

$$\delta\boldsymbol{p} = E\left\{\Delta\boldsymbol{p}\right\} = \boldsymbol{B}_3^{\circ-1}\left(\delta\boldsymbol{\varphi}_2 - \boldsymbol{c}_p\right) \tag{9.54}$$

where $\boldsymbol{c}_p$ is a column vector which contains the diagonal elements of the error covariance matrix $\boldsymbol{C}_p$ in (9.44).

A direct solution for this bias problem is to estimate $\delta\boldsymbol{p}$ and subtract it from $\hat{\boldsymbol{p}}$. Note that, since the exact value of $\delta\boldsymbol{p}$ cannot be computed as it is a function of $\boldsymbol{p}$, $\delta\boldsymbol{p}$ can only be approximated using $\hat{\boldsymbol{p}}$. However, this approach requires that the TDOA noise covariance matrix $\boldsymbol{Q}$ must be exactly known. However, it is often the case in practice that only the structure of $\boldsymbol{Q}$ is available due to the lack of knowledge of TDOA noise power which has to be estimated separately.

9.5 Bias compensation techniques

9.5.1 Augmented solution with quadratic constraint

The bias problem of the three-stage least-squares solution can be alleviated by formulating an augmented solution equation and imposing a quadratic constraint [114]. The objective function for the least-squares solution $\hat{\boldsymbol{\varphi}}_1$ in (9.22) is

$$\epsilon = (\boldsymbol{h}_1 - \mathbf{G}_1\boldsymbol{\varphi}_1)^T \boldsymbol{W}_1 (\boldsymbol{h}_1 - \mathbf{G}_1\boldsymbol{\varphi}_1), \tag{9.55}$$

which can be equivalently rewritten as

$$\epsilon = \boldsymbol{v}^T \boldsymbol{A}^T \boldsymbol{W}_1 \boldsymbol{A} \boldsymbol{v} \tag{9.56}$$

where the augmented matrix $\boldsymbol{A}$ and vector $\boldsymbol{v}$ are defined by

$$\boldsymbol{A} = [-\mathbf{G}_1, \boldsymbol{h}_1], \tag{9.57a}$$

$$\boldsymbol{v} = \left[\boldsymbol{\varphi}_1^T, 1\right]^T. \tag{9.57b}$$

The augmented matrix $\boldsymbol{A}$ can be decomposed into the sum of its true noise-free version $\boldsymbol{A}^\circ$ and a noise matrix $\Delta\boldsymbol{A}$:

$$\boldsymbol{A} = \boldsymbol{A}^\circ + \Delta\boldsymbol{A} \tag{9.58}$$

where

$$\Delta\boldsymbol{A} = 2\left[\mathbf{0}_{(N-1)\times D}, \boldsymbol{n}, \bar{\boldsymbol{B}}_1\boldsymbol{n}\right], \tag{9.59a}$$

$$\bar{\boldsymbol{B}}_1 = \mathrm{diag}(d_{21}, d_{31}, \ldots, d_{N1}). \tag{9.59b}$$

Substituting (9.58) into (9.56) and taking the expectation of ϵ yields the averaged objective function

$$E\{\epsilon\} = \boldsymbol{v}^T \boldsymbol{A}^{\circ T} \boldsymbol{W}_1 \boldsymbol{A}^\circ \boldsymbol{v} + \boldsymbol{v}^T E\{\Delta\boldsymbol{A}^T \boldsymbol{W}_1 \Delta\boldsymbol{A}\}\boldsymbol{v}. \tag{9.60}$$

Both terms in the right-hand side of (9.60) are quadratic forms and therefore nonnegative functions of $\boldsymbol{v}$. The first term $\boldsymbol{v}^T \boldsymbol{A}^{\circ T} \boldsymbol{W}_1 \boldsymbol{A}^\circ \boldsymbol{v}$ is the ideal cost function which has the minimum value of zero at the true $\boldsymbol{v}$ because $\boldsymbol{A}^\circ\boldsymbol{v} = \boldsymbol{h}_1^\circ - \mathbf{G}_1^\circ\boldsymbol{\varphi}_1 = \mathbf{0}$ for the true $\boldsymbol{\varphi}_1$. If $E\{\epsilon\}$ is minimized with respect to $\boldsymbol{v}$, the second term $\boldsymbol{v}^T E\{\Delta\boldsymbol{A}^T \boldsymbol{W}_1 \Delta\boldsymbol{A}\}\boldsymbol{v}$ steers the solution away from the true $\boldsymbol{v}$, thus resulting in a bias in the estimate of $\boldsymbol{v}$.

This problem can be overcome by imposing a constraint such that the second term of (9.60) is a constant when minimizing ϵ, which leads to

$$\min_{\boldsymbol{v}} \left\{\boldsymbol{v}^T\boldsymbol{A}^T\boldsymbol{W}_1\boldsymbol{A}\boldsymbol{v}\right\} \quad \text{subject to} \quad \boldsymbol{v}^T\boldsymbol{\Omega}\boldsymbol{v}=k \tag{9.61}$$

where $\boldsymbol{\Omega}=E\{\Delta\boldsymbol{A}^T\boldsymbol{W}_1\Delta\boldsymbol{A}\}$. Note that the constant k only affects the scaling of $\boldsymbol{v}$ and thus its value is not important. This constrained minimization problem can be solved via the Lagrange multiplier method

$$\min_{\boldsymbol{v}} \left\{\boldsymbol{v}^T\boldsymbol{A}^T\boldsymbol{W}_1\boldsymbol{A}\boldsymbol{v}+\lambda(k-\boldsymbol{v}^T\boldsymbol{\Omega}\boldsymbol{v})\right\} \tag{9.62}$$

where λ is the Lagrange multiplier. Taking the derivative of the argument inside the minimization in (9.62) with respect to $\boldsymbol{v}$ and setting it to zero yields

$$(\boldsymbol{A}^T\boldsymbol{W}_1\boldsymbol{A})\boldsymbol{v}=\lambda\boldsymbol{\Omega}\boldsymbol{v}. \tag{9.63}$$

Multiplying both sides of (9.63) with $\boldsymbol{v}^T$ from to the left and noting from (9.61) that $\boldsymbol{v}^T\boldsymbol{\Omega}\boldsymbol{v}=k$, the cost function to be minimized in (9.61) becomes λk. Since k is a constant, λ must be minimized. As a result, the solution of $\boldsymbol{v}$, denoted as $\hat{\boldsymbol{v}}$, is the generalized eigenvector of the pair $(\boldsymbol{A}^T\boldsymbol{W}_1\boldsymbol{A},\boldsymbol{\Omega})$ which corresponds to the smallest generalized eigenvalue $\lambda_{\min}$. Although generalized eigen-decomposition techniques can be used to find $\hat{\boldsymbol{v}}$, an explicit solution for $\hat{\boldsymbol{v}}$ is available as shown now.

Using (9.59), the constraint matrix $\boldsymbol{\Omega}$ becomes

$$\boldsymbol{\Omega}=E\{\Delta\boldsymbol{A}^T\boldsymbol{W}_1\Delta\boldsymbol{A}\}=\begin{bmatrix}\boldsymbol{0}_{D\times D} & \boldsymbol{0}_{2\times D}\\ \boldsymbol{0}_{D\times 2} & \bar{\boldsymbol{\Omega}}\end{bmatrix} \tag{9.64}$$

where

$$\bar{\boldsymbol{\Omega}}=4\begin{bmatrix}\mathrm{trace}(\boldsymbol{W}_1\boldsymbol{Q}) & \mathrm{trace}(\boldsymbol{W}_1\bar{\boldsymbol{B}}_1\boldsymbol{Q})\\ \mathrm{trace}(\bar{\boldsymbol{B}}_1\boldsymbol{W}_1\boldsymbol{Q}) & \mathrm{trace}(\bar{\boldsymbol{B}}_1\boldsymbol{W}_1\bar{\boldsymbol{B}}_1\boldsymbol{Q})\end{bmatrix}. \tag{9.65}$$

Note that $\bar{\boldsymbol{\Omega}}$ is dependent on $\bar{\boldsymbol{B}}_1$ which is a function of the unknown true TDOAs. Thus, the noisy TDOA measurements are used instead for the computation of $\bar{\boldsymbol{B}}_1$ and $\bar{\boldsymbol{\Omega}}$.

The vector $\boldsymbol{v}$ and matrix $\boldsymbol{A}^T\boldsymbol{W}_1\boldsymbol{A}$ can be partitioned into

$$\boldsymbol{v}=\begin{bmatrix}\boldsymbol{v}_1\\ \boldsymbol{v}_2\end{bmatrix}, \quad \boldsymbol{A}^T\boldsymbol{W}_1\boldsymbol{A}=\begin{bmatrix}\boldsymbol{A}_{11} & \boldsymbol{A}_{12}\\ \boldsymbol{A}_{12}^T & \boldsymbol{A}_{22}\end{bmatrix}, \tag{9.66}$$

where $\boldsymbol{v}_1$ and $\boldsymbol{v}_2$ are column vectors with dimensions $D \times 1$ and 2×1, respectively, and the dimensions of sub-matrices $\boldsymbol{A}_{11}$, $\boldsymbol{A}_{12}$ and $\boldsymbol{A}_{22}$ are $D \times D$, $2 \times D$ and 2×2, respectively.

Substituting (9.64)–(9.66) into (9.63) yields

$$\begin{bmatrix} \boldsymbol{A}_{11}\boldsymbol{v}_1 + \boldsymbol{A}_{12}\boldsymbol{v}_2 \\ \boldsymbol{A}_{12}^T\boldsymbol{v}_1 + \boldsymbol{A}_{22}\boldsymbol{v}_2 \end{bmatrix} = \begin{bmatrix} \boldsymbol{0} \\ \lambda\bar{\boldsymbol{\Omega}}\boldsymbol{v}_2 \end{bmatrix} \tag{9.67}$$

whence we obtain

$$\boldsymbol{v}_1 = -\boldsymbol{A}_{11}^{-1}\boldsymbol{A}_{12}\boldsymbol{v}_2 \tag{9.68}$$

and

$$(\boldsymbol{A}_{22} - \boldsymbol{A}_{12}^T\boldsymbol{A}_{11}^{-1}\boldsymbol{A}_{12})\boldsymbol{v}_2 = \lambda\bar{\boldsymbol{\Omega}}\boldsymbol{v}_2. \tag{9.69}$$

The positive definiteness of $\bar{\boldsymbol{\Omega}}$ ensures a unique solution of $\boldsymbol{v}_2$. To find $\boldsymbol{v}_2$, we rewrite (9.69) as

$$\boldsymbol{D}\boldsymbol{v}_2 = \lambda\boldsymbol{v}_2 \tag{9.70}$$

where

$$\boldsymbol{D} = \bar{\boldsymbol{\Omega}}^{-1}(\boldsymbol{A}_{22} - \boldsymbol{A}_{12}^T\boldsymbol{A}_{11}^{-1}\boldsymbol{A}_{12}) = \begin{bmatrix} D_{11} & D_{12} \\ D_{21} & D_{22} \end{bmatrix}. \tag{9.71}$$

Noting that the eigenvalue λ satisfies $|\boldsymbol{D} - \lambda\boldsymbol{I}| = 0$, the following quadratic equation can be obtained:

$$\lambda^2 - (D_{11} + D_{22})\lambda + D_{11}D_{22} - D_{12}D_{21} = 0 \tag{9.72}$$

with the smaller root given by

$$\lambda_{\min} = \frac{1}{2}\left(D_{11} + D_{22} - \left((D_{11} - D_{22})^2 + 4D_{12}D_{21}\right)^{1/2}\right). \tag{9.73}$$

Substituting $\lambda_{\min}$ into (9.70) and normalizing the second element of $\boldsymbol{v}_2$ to unity, we obtain the solution for $\boldsymbol{v}_2$ as

$$\hat{\boldsymbol{v}}_2 = \begin{bmatrix} \left(\dfrac{\lambda_{\min} - D_{22}}{D_{21}}\right) \\ 1 \end{bmatrix}. \tag{9.74}$$

Substituting $\hat{\boldsymbol{v}}_2$ into (9.68), we obtain the solution for $\boldsymbol{v}_1$ as

$$\hat{\boldsymbol{v}}_1 = -\boldsymbol{A}_{11}^{-1}\boldsymbol{A}_{12}\hat{\boldsymbol{v}}_2. \tag{9.75}$$

Thus, the solution for $\boldsymbol{\varphi}_1$ can be obtained from $\hat{\boldsymbol{v}}_1$ and $\hat{\boldsymbol{v}}_2$ as

$$\hat{\boldsymbol{\varphi}}_1 = \begin{bmatrix} \hat{\boldsymbol{v}}_1 \\ \hat{\boldsymbol{v}}_2(1) \end{bmatrix} \tag{9.76}$$

where $\hat{\boldsymbol{v}}_2(1)$ denotes the first element of $\hat{\boldsymbol{v}}_2$. Since the bias problem has already been resolved in the estimate $\hat{\boldsymbol{\varphi}}_1$, the estimates $\hat{\boldsymbol{\varphi}}_2$ and $\hat{\boldsymbol{p}}$ are subsequently calculated using (9.36) and (9.41) without any further bias compensation.

9.5.2 Instrumental-variable based solution

In this section, another bias reduction method [116] is presented based on the framework of IV estimation. Compared to the augmented solution with quadratic constraint, this method has a lower computational complexity while producing an almost identical performance both in terms of bias and mean-squared error.

Recall from Section 9.4 that the bias of the three-stage least-squares estimator arises from the correlation between the measurement matrix $\boldsymbol{G}_1$ and pseudolinear noise $\boldsymbol{\eta}_1$. This correlation can be overcome by modifying the normal equations of the weighted least-squares estimation in (9.22)

$$(\boldsymbol{G}_1^T \boldsymbol{W}_1 \boldsymbol{G}_1)\hat{\boldsymbol{\varphi}}_1 = \boldsymbol{G}_1^T \boldsymbol{W}_1 \boldsymbol{h}_1 \tag{9.77}$$

to

$$(\boldsymbol{F}_1^T \boldsymbol{W}_1 \boldsymbol{G}_1)\hat{\boldsymbol{\varphi}}_1^{\mathrm{IV}} = \boldsymbol{F}_1^T \boldsymbol{W}_1 \boldsymbol{h}_1 \tag{9.78}$$

using an IV matrix $\boldsymbol{F}_1$, which leads to an IV estimate given by

$$\hat{\boldsymbol{\varphi}}_1^{\mathrm{IV}} = (\boldsymbol{F}_1^T \boldsymbol{W}_1 \boldsymbol{G}_1)^{-1} \boldsymbol{F}_1^T \boldsymbol{W}_1 \boldsymbol{h}_1. \tag{9.79}$$

In theory, this IV-based estimate is asymptotically unbiased, i.e., $E\{\hat{\boldsymbol{\varphi}}_1^{\mathrm{IV}} - \boldsymbol{\varphi}_1\} \to \mathbf{0}$ as $N \to \infty$, if the IV matrix $\boldsymbol{F}_1$ is selected so that $E\left\{\frac{\boldsymbol{F}_1^T \boldsymbol{W}_1 \boldsymbol{G}_1}{N-1}\right\}$ is nonsingular and $E\left\{\frac{\boldsymbol{F}_1^T \boldsymbol{W}_1 \boldsymbol{h}_1}{N-1}\right\} = \mathbf{0}$ as $N \to \infty$ [80].

Since the noise-free measurement matrix $\boldsymbol{G}_1^\circ$, which is the optimal choice for the IV matrix $\boldsymbol{F}_1$, is a function of the unknown target position $\boldsymbol{p}$, the target position information extracted from $\hat{\boldsymbol{\varphi}}_1$ in (9.22), i.e.,

$\hat{\boldsymbol{\varphi}}_1(1:D)$, can be used instead to construct the IV matrix as

$$\boldsymbol{F}_1 = -2 \begin{bmatrix} (\boldsymbol{r}_2 - \boldsymbol{r}_1)^T & \hat{d}_{21}^{\mathrm{IV}} \\ (\boldsymbol{r}_3 - \boldsymbol{r}_1)^T & \hat{d}_{31}^{\mathrm{IV}} \\ \vdots & \vdots \\ (\boldsymbol{r}_N - \boldsymbol{r}_1)^T & \hat{d}_{N1}^{\mathrm{IV}} \end{bmatrix} \tag{9.80}$$

where $\hat{d}_{i1}^{\mathrm{IV}} = \|\hat{\boldsymbol{\varphi}}_1(1:D) - \boldsymbol{r}_i\| - \|\hat{\boldsymbol{\varphi}}_1(1:D) - \boldsymbol{r}_1\|$.

Since the bias of $\hat{\boldsymbol{\varphi}}_1$ is already resolved by the IV-based solution $\hat{\boldsymbol{\varphi}}_1^{\mathrm{IV}}$, the estimates in (9.36) and (9.41) for $\hat{\boldsymbol{\varphi}}_2$ and $\hat{\boldsymbol{p}}$, respectively, are now performed with $\hat{\boldsymbol{\varphi}}_1$ replaced by $\hat{\boldsymbol{\varphi}}_1^{\mathrm{IV}}$ and no further bias compensation procedure is required.

For a finite number of receivers, this IV-based method is not strictly unbiased. This notwithstanding, it is capable of significantly reducing the bias of the original three-stage least-squares solution. In the next section, the effectiveness of the IV-based solution in terms of bias reduction is demonstrated via numerical examples where it is observed to exhibit a performance on a par with the augmented solution presented in Section 9.5.1. More importantly, the IV-based solution is computationally more efficient than the augmented solution as shown analytically in Table 9.1 and demonstrated numerically in Section 9.6.

9.6 Algorithm performance and comparison

This section presents simulation examples to evaluate the performance of the algorithms covered in Sections 9.2–9.5 in terms of bias and mean-squared error. The MSE and bias of target position estimates are defined by

$$\mathrm{MSE} = \frac{1}{N_{\mathrm{MC}}} \sum_{l=1}^{N_{\mathrm{MC}}} \left\| \hat{\boldsymbol{p}}^{\langle l \rangle} - \boldsymbol{p} \right\|^2, \tag{9.81a}$$

$$\mathrm{bias} = \left\| \frac{1}{N_{\mathrm{MC}}} \sum_{l=1}^{N_{\mathrm{MC}}} \left(\hat{\boldsymbol{p}}^{\langle l \rangle} - \boldsymbol{p} \right) \right\|, \tag{9.81b}$$

respectively, where N_{MC} is the number of Monte Carlo runs and $\hat{\boldsymbol{p}}^{\langle l \rangle}$ is the estimate of $\boldsymbol{p}$ obtained in the lth Monte Carlo run. The Gauss–Newton iteration is used to implement the MLE, which is initialized by the true target position and halted after 10 iterations. The sum of the diagonal elements of the CRLB matrix $\boldsymbol{C}_{\mathrm{CRLB}}$ in (9.11) is computed and used as a

Table 9.1 Comparison of computational complexities†.

Algorithm	Computation	
Three-stage least-squares solution (presented in Section 9.3)	Multiplication	$(2D+6)(N-1)^2+(2D^2+7D+6)(N-1)+2(D+1)^2$
	Division	$N-1$
	Square root	$N-1$
	Matrix inversion	$2\,\mathrm{inv}_{(D+1)\times(D+1)}$
Augmented solution with quadratic constraint (presented in Section 9.5.1)	Multiplication	$4(N-1)^3+(2D+10)(N-1)^2+(2D^2+9D+9)(N-1)+(4D^2+8D+40)$
	Division	$N+2D+3$
	Square root	$N+1$
	Matrix inversion	$4\,\mathrm{inv}_{D\times D}+2\,\mathrm{inv}_{2\times2}$
IV-based solution (presented in Section 9.5.2)	Multiplication	$(3D+8)(N-1)^2+(3D^2+11D+8)(N-1)+(3D^2+7D+3)$
	Division	$N-1$
	Square root	$2N-1$
	Matrix inversion	$3\,\mathrm{inv}_{(D+1)\times(D+1)}$

† Only the complexity for finding $\boldsymbol{\varphi}_1$ is considered as the procedures of solving $\boldsymbol{\varphi}_2$ and $\boldsymbol{p}$ are identical for all the methods. Here, $\mathrm{inv}_{K\times K}$ denotes an inversion of the $K\times K$ matrix. The computations for the terms which can be precalculated such as $\mathbf{Q}^{-1}$ and $\boldsymbol{r}_i^T\boldsymbol{r}$ are excluded.

benchmark for the MSE performance of the algorithms. Two scenarios are considered in the simulations as described below.

Scenario 1 (fixed geometry): A passive multistatic system with five receivers is considered, where the first receiver (the reference receiver) is located at the origin $[0,0]^T$ km while the remaining receivers are uniformly located along a circle of radius 10 km, i.e., $\boldsymbol{r}_i = \left[10\cos\left(\frac{2\pi}{N-1}(i-2)\right), 10\sin\left(\frac{2\pi}{N-1}(i-2)\right)\right]^T$ km for $i=2,3,4$ and 5. The target is located at $\boldsymbol{p}=\left[100\cos\left(\frac{2\pi}{32}\right), 100\sin\left(\frac{2\pi}{32}\right)\right]^T$ km. The TDOA noise covariance matrix is set to $\mathbf{Q}=\sigma^2(\boldsymbol{I}+\mathbf{1})/2$. A total of $N_{\mathrm{MC}}=10\,000$ Monte Carlo runs are carried out.

Scenario 2 (random geometry): Random localization geometries are considered, where 2000 geometries are randomly generated based on the uniform distribution. Eight receivers are allocated randomly within 10 km from the origin, while the target is placed randomly with a distance from the origin in between 10 km and 60 km. The MSE and bias are first obtained over 1000 Monte Carlo runs for each geometry and then averaged over all 2000 geometries.

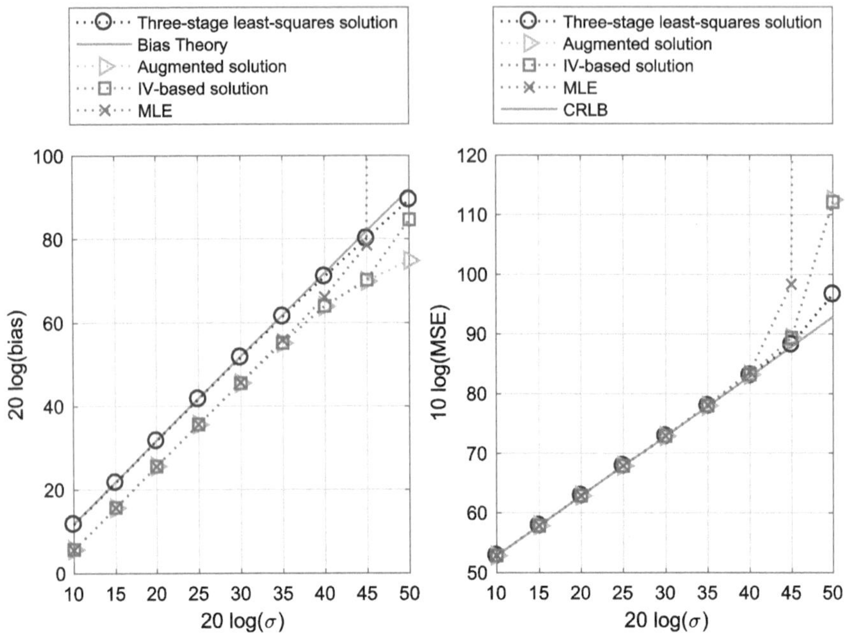

Figure 9.2 Performance comparison between the three-stage least-squares solution, augmented solution, IV-based solution, and MLE for Scenario 1.

Figs. 9.2 and 9.3 show the bias and MSE performance of the three-stage least-squares solution, augmented solution, IV-based solution, and MLE for Scenarios 1 and 2, respectively. Here, $20\log(\text{bias})$, $10\log(\text{MSE})$ and $20\log(\sigma)$ have the same unit of $20\log(\text{meter})$. We observe that the three-stage least-squares solution suffers from a severe bias which is much larger than that of the MLE. In addition, the bias of the three-stage least-squares solution agrees with the theoretical bias value presented in Section 9.4. In contrast, the augmented solution with quadratic constraint and the IV-based solution yield an almost identical bias performance which is much smaller than the bias of the three-stage least-squares solution. Moreover, the bias performance of these two methods is very close to that of the MLE, thus demonstrating their capability to achieve a bias performance similar to optimal nonlinear estimation. In addition, we also observe from Figs. 9.2 and 9.3 that these two methods exhibit MSEs almost identical to that of the MLE and their MSEs closely achieve the CRLB before the threshold effect occurs at $20\log(\sigma) = 45$ for Scenario 1 and at $20\log(\sigma) = 20$ for Scenario 2.

To numerically compare the complexity of the algorithms, all the algorithms were implemented in MATLAB and executed on the same hard-

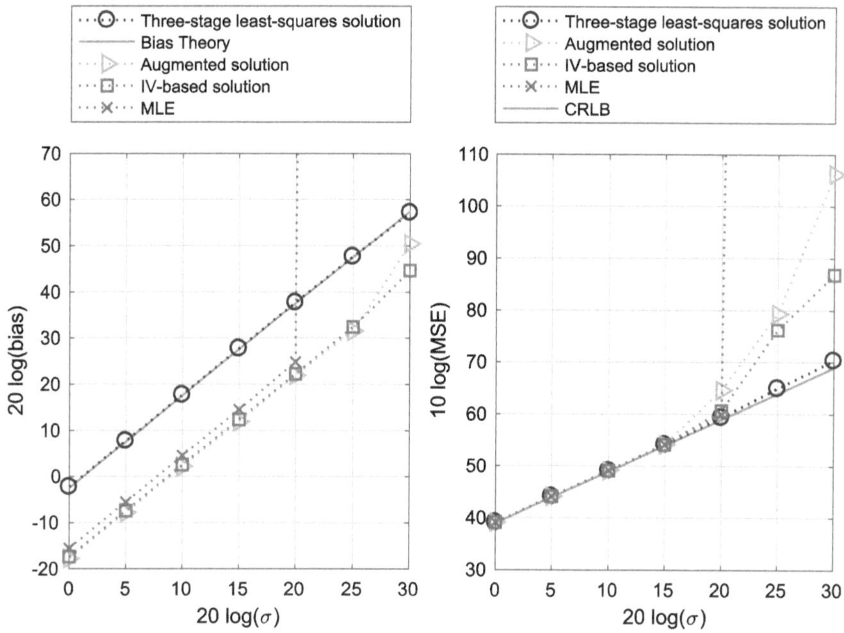

Figure 9.3 Performance comparison between the three-stage least-squares solution, augmented solution, IV-based solution, and MLE for Scenario 2.

Table 9.2 Averaged runtime.

Algorithm	**Runtime**[†]
Three-stage least-squares solution	1
IV-based solution	1.24
Augmented solution	1.94
MLE	3.61

[†] Normalized by the runtime of the three-stage least-squares solution.

ware platform. The averaged runtimes of the algorithms are reported in Table 9.2, where the averaged runtimes are normalized by the averaged runtime of the three-stage least-squares solution. We observe that, although producing a comparable performance, the IV-based solution is computationally more efficient than the augmented solution and MLE. Specifically, the IV-based solution is only 24% slower than the three-stage least-squares solution due to the additional IV estimation, while the augmented solution and MLE require 94% and 261% longer runtimes compared to the three-stage least-squares solution, respectively.

9.7 Summary

This chapter has considered the problem of multistatic target localization using passive TDOA measurements. The iterative MLE solutions are not only computationally burdensome but also susceptible to convergence to local minima or divergence due to the nonlinear nature of the TDOA measurement equation. Alternatively, the nonlinear TDOA measurements can be algebraically rearranged into a set of linear equations by introducing a nuisance parameter, thus allowing closed-form linear least-squares to be used. This approach requires the constraint relationship between the target position and nuisance parameter to be taken into consideration to construct efficient estimators, while the correlation between the measurement matrix and pseudolinear noise vector must be handled to prevent estimation bias. To address these requirements, three closed-form solutions were presented; viz., the three-stage least-squares solution, the augmented solution with quadratic constraint, and the IV-based solution. Simulation examples were presented to demonstrate the efficacy of these algorithms.

Bibliography

[1] M.I. Skolnik, Radar Handbook, third ed., McGraw-Hill, New York, NY, 2008.

[2] M.A. Richards, Fundamentals of Radar Signal Processing, McGraw-Hill, New York, 2005.

[3] P. Tait, Introduction to Radar Target Recognition, Institution of Electrical Engineers, London, UK, 2005.

[4] D.R. Wehner, High-Resolution Radar, Artech House, Boston, MA, 1995.

[5] V.C. Chen, M. Martorella, Inverse Synthetic Aperture Radar Imaging: Principles, Algorithms and Applications, SciTech Publishing, Edison, NJ, 2014.

[6] A. Farina, F.A. Studer, Radar Data Processing: Vol. II – Advanced Topics and Applications, Research Studies Press, UK, 1986.

[7] N.J. Willis, H.D. Griffiths (Eds.), Advances in Bistatic Radar, SciTech Publishing, Raleigh, NC, 2007.

[8] F. Gini, A. De Maio, L. Patton (Eds.), Waveform Design and Diversity for Advanced Radar Systems, Institution of Engineering and Technology, London, UK, 2012.

[9] M.G. Amin (Ed.), Radar for Indoor Monitoring: Detection, Classification, and Assessment, CRC Press, Boca Raton, FL, 2017.

[10] M.G. Amin (Ed.), Through-the-Wall Radar Imaging, CRC Press, Boca Raton, FL, 2017.

[11] R.J. Doviak, D.S. Zrnic, Doppler Radar and Weather Observations, second ed., Academic Press, San Diego, 1993.

[12] V.N. Bringi, V. Chandrasekar, Polarimetric Doppler Weather Radar: Principles and Applications, Cambridge University Press, Cambridge, UK, 2001.

[13] N.J. Willis, Introduction, in: N.J. Willis, H.D. Griffiths (Eds.), Advances in Bistatic Radar, SciTech Publishing, Raleigh, NC, 2007, pp. 3–9, chap. 1.

[14] N.J. Willis, Bistatic Radar, SciTech Publishing, Raleigh, NC, 2005.

[15] A.M. Haimovich, R.S. Blum, L.J. Cimini, MIMO radar with widely separated antennas, IEEE Signal Processing Magazine 25 (1) (2008) 116–129.

[16] J. Li, P. Stoica, MIMO Radar Signal Processing, John Wiley and Sons, Hoboken, New Jersey, 2009.

[17] H.D. Griffiths, C.J. Baker, An Introduction to Passive Radar, Artech House, Norwood, MA, 2017.

[18] M.C. Wicks, E.L. Mokole, S.D. Blunt, R.S. Schneible, V.J. Amuso (Eds.), Principles of Waveform Diversity and Design, SciTech Publishing, Raleigh, NC, 2010.

[19] H.L. Van Trees, Detection, Estimation, and Modulation Theory, Part III: Radar-Sonar Signal Processing and Gaussian Signals in Noise, John Wiley and Sons, New York, 2001.

[20] T. Tsao, M. Slamani, P. Varshney, D. Weiner, H. Schwarzlander, S. Borek, Ambiguity function for a bistatic radar, IEEE Transactions on Aerospace and Electronic Systems 33 (3) (1997) 1041–1051.

[21] I. Bradaric, G.T. Capraro, M.C. Wicks, Waveform diversity and signal processing strategies in multistatic radar systems, in: M.C. Wicks, E.L. Mokole, S.D. Blunt, R.S. Schneible, V.J. Amuso (Eds.), Principles of Waveform Diversity and Design, SciTech Publishing, Raleigh, NC, 2010, pp. 286–304, chap. 16.

[22] S.M. Kay, Fundamentals of Statistical Signal Processing: Estimation Theory, Prentice-Hall, Englewood Cliffs, NJ, 1993.
[23] Y. Bar-Shalom, X. Li Rong, T. Kirubarajan, Estimation With Applications to Tracking and Navigation: Theory, Algorithms and Software, John Wiley and Sons, New York, 2001.
[24] S. Blackman, R. Popoli, Design and Analysis of Modern Tracking Systems, Artech House, Boston, MA, 1999.
[25] S.D. Howard, S. Suvorova, W. Moran, Waveform libraries for radar tracking applications, in: Proc. International Waveform Diversity Design Conference, Edinburgh, UK, 2004, pp. 1–5.
[26] S. Suvorova, S.D. Howard, W. Moran, R.J. Evans, Waveform libraries for radar tracking applications: maneuvering targets, in: Proc. 40th Annual Conference on Information Sciences and Systems, Princeton, NJ, USA, 2006, pp. 1424–1428.
[27] A. Leshem, O. Naparstek, A. Nehorai, Information theoretic adaptive radar waveform design for multiple extended targets, IEEE Journal of Selected Topics in Signal Processing 1 (1) (2007) 42–55.
[28] D.J. Kershaw, R.J. Evans, Optimal waveform selection for tracking systems, IEEE Transactions on Information Theory 40 (5) (1994) 1536–1550.
[29] D.J. Kershaw, R.J. Evans, Waveform selective probabilistic data association, IEEE Transactions on Aerospace and Electronic Systems 33 (4) (1997) 1180–1188.
[30] S.P. Sira, A. Papandreou-Suppappola, D. Morrell, Dynamic configuration of time-varying waveforms for agile sensing and tracking in clutter, IEEE Transactions on Signal Processing 55 (7) (2007) 3207–3217.
[31] S.P. Sira, A. Papandreou-Suppappola, D. Morrell, Advances in Waveform-Agile Sensing for Tracking, Synthesis Lectures on Algorithms and Software in Engineering, Morgan and Claypool, San Rafael, CA, 2008.
[32] S.P. Sira, Y. Li, A. Papandreou-Suppappola, D. Morrell, D. Cochran, M. Rangaswamy, Waveform-agile sensing for tracking, IEEE Signal Processing Magazine 26 (1) (2009) 53–64.
[33] N.H. Nguyen, K. Dogancay, L.M. Davis, Joint transmitter waveform and receiver path optimization for target tracking by multistatic radar system, in: Proc. EEE Workshop on Statistical Signal Processing (SSP), Gold Coast, Australia, 2014, pp. 444–447.
[34] N.H. Nguyen, K. Dogancay, L.M. Davis, Adaptive waveform selection and target tracking by wideband multistatic radar/sonar systems, in: Proc. European Signal Processing Conference (EUSIPCO), Lisbon, Portugal, 2014, pp. 1910–1914.
[35] Y. Bar-Shalom, W.D. Blair, Multitarget-Multisensor Tracking: Application and Advances, vol. 3, Artech House, Norwood, MA, 2000.
[36] S.-M. Hong, R.J. Evans, H.-S. Shin, Optimization of waveform and detection threshold for range and range-rate tracking in clutter, IEEE Transactions on Aerospace and Electronic Systems 41 (1) (2005) 17–33.
[37] M.S. Greco, P. Stinco, F. Gini, A. Farina, Cramer–Rao bounds and selection of bistatic channels for multistatic radar systems, IEEE Transactions on Aerospace and Electronic Systems 47 (4) (2011) 2934–2948.
[38] P. Stinco, M.S. Greco, F. Gini, A. Farina, Posterior Cramer–Rao lower bounds for passive bistatic radar tracking with uncertain target measurements, Signal Processing 93 (12) (2013) 3528–3540.
[39] T. Tsao, M. Slamani, P. Varshney, D. Weiner, H. Schwarzlander, S. Borek, Ambiguity function for a bistatic radar, IEEE Transactions on Aerospace and Electronic Systems 33 (3) (1997) 1041–1051.

[40] A. Dogandzic, A. Nehorai, Cramer–Rao bounds for estimating range, velocity, and direction with an active array, IEEE Transactions on Signal Processing 49 (6) (2001) 1122–1137.

[41] A. Dogandzic, A. Nehorai, Estimating range, velocity, and direction with a radar array, in: Proc. IEEE International Conference on Acoustics, Speech and Signal Processing (ICASSP), Cambridge, MA, vol. 5, 1999, pp. 2773–2776.

[42] X. Li Rong, V.P. Jilkov, Survey of maneuvering target tracking. Part I. Dynamic models, IEEE Transactions on Aerospace and Electronic Systems 39 (4) (2003) 1333–1364.

[43] E. Mazor, A. Averbuch, Y. Bar-Shalom, J. Dayan, Interacting multiple model methods in target tracking: a survey, IEEE Transactions on Aerospace and Electronic Systems 34 (1) (1998) 103–123.

[44] C.O. Savage, B. Moran, Waveform selection for maneuvering targets within an IMM framework, IEEE Transactions on Aerospace and Electronic Systems 43 (3) (2007) 1205–1214.

[45] Y. Bar-Shalom, X. Li, Multitarget-Multisensor Tracking: Principles and Techniques, YBS Publishing, Storrs, CT, 1995.

[46] S. Blackman, Multi-Target Tracking With Radar Applications, Artech House, Norwood, MA, 1986.

[47] A. Farina, F.A. Studer, Radar Data Processing: Vol. II – Advanced Topics and Applications, Research Studies Press, UK, 1986.

[48] T. Fortmann, Y. Bar-Shalom, M. Scheffe, S. Gelfand, Detection thresholds for tracking in clutter–a connection between estimation and signal processing, IEEE Transactions on Automatic Control 30 (3) (1985) 221–229.

[49] D.J. Kershaw, R.J. Evans, A contribution to performance prediction for probabilistic data association tracking filters, IEEE Transactions on Aerospace and Electronic Systems 32 (3) (1996) 1143–1148.

[50] Y. Bar-Shalom, T.E. Fortmann, Tracking and Data Association, Academic Press, San Diego, CA, 1988.

[51] A. Farina, Tracking function in bistatic and multistatic radar systems, IEE Proceedings F (Communications, Radar and Signal Processing) 133 (7) (1986) 630–637.

[52] M.S. Mahmoud, H.M. Khalid, Distributed Kalman filtering: a bibliographic review, IET Control Theory & Applications 7 (4) (2013) 483–501.

[53] F.S. Cattivelli, A.H. Sayed, Diffusion strategies for distributed Kalman filtering and smoothing, IEEE Transactions on Automatic Control 55 (9) (2010) 2069–2084.

[54] R. Carli, A. Chiuso, L. Schenato, S. Zampieri, Distributed Kalman filtering based on consensus strategies, IEEE Journal on Selected Areas in Communications 26 (4) (2008) 622–633.

[55] Y. Oshman, P. Davidson, Optimization of observer trajectories for bearings-only target localization, IEEE Transactions on Aerospace and Electronic Systems 35 (3) (1999) 892–902.

[56] K. Dogancay, Online optimization of receiver trajectories for scan-based emitter localization, IEEE Transactions on Aerospace and Electronic Systems 43 (3) (2007) 1117–1125.

[57] J. Ousingsawat, M.E. Campbell, Optimal cooperative reconnaissance using multiple vehicles, Journal of Guidance, Control, and Dynamics 30 (1) (2007) 122–132.

[58] K. Dogancay, UAV path planning for passive emitter localization, IEEE Transactions on Aerospace and Electronic Systems 48 (2) (2012) 1150–1166.

[59] K. Dogancay, H. Hmam, Optimal angular sensor separation for AOA localization, Signal Processing 88 (5) (2008) 1248–1260.
[60] A.N. Bishop, B. Fidan, B.D.O. Anderson, K. Dogancay, P.N. Pathirana, Optimality analysis of sensor-target localization geometries, Automatica 46 (3) (2010) 479–492.
[61] S. Zhao, B.M. Chen, T.H. Lee, Optimal sensor placement for target localisation and tracking in 2D and 3D, International Journal of Control 86 (10) (2013) 1687–1704.
[62] K. Dogancay, H. Hmam, On optimal sensor placement for time-difference-of-arrival localization utilizing uncertainty minimization, in: Proc. European Signal Processing Conference (EUSIPCO), Glasgow, Scotland, 2009, pp. 1136–1140.
[63] K.W.K. Lui, H.C. So, A study of two-dimensional sensor placement using time-difference-of-arrival measurements, Digital Signal Processing 19 (4) (2009) 650–659.
[64] N.H. Nguyen, K. Dogancay, Optimal sensor placement for Doppler shift target localization, in: Proc. IEEE Radar Conference (RadarCon), Arlington, VA, USA, 2015, pp. 1677–1682.
[65] N.H. Nguyen, K. Dogancay, Optimal sensor-target geometries for Doppler-shift target localization, in: Proc. European Signal Processing Conference (EUSIPCO), Nice, France, 2015, pp. 180–184.
[66] D. Ucinski, Optimal Measurement Methods for Distributed Parameter System Identification, CRC Press, Boca Raton, FL, 2004.
[67] H. Godrich, A.M. Haimovich, R.S. Blum, Target localization accuracy gain in MIMO radar-based systems, IEEE Transactions on Information Theory 56 (6) (2010) 2783–2803.
[68] L. Rui, K.C. Ho, Elliptic localization: performance study and optimum receiver placement, IEEE Transactions on Signal Processing 62 (18) (2014) 4673–4688.
[69] A. Beck, P. Stoica, J. Li, Exact and approximate solutions of source localization problems, IEEE Transactions on Signal Processing 56 (5) (2008) 1770–1778.
[70] J. Shen, A.F. Molisch, J. Salmi, Accurate passive location estimation using TOA measurements, IEEE Transactions on Wireless Communications 11 (6) (2012) 2182–2192.
[71] P.S. Bullen, Handbook of Means and Their Inequalities, Springer, Netherlands, Dordrecht, 2003.
[72] D.E. Goldberg, Genetic Algorithms in Search, Optimization, and Machine Learning, Addison-Wesley, 1989.
[73] J.A. Snyman, Practical Mathematical Optimization: An Introduction to Basic Optimization Theory and Classical and New Gradient-Based Algorithms, Springer, New York, NY, 2005.
[74] D.J. Torrieri, Statistical theory of passive location systems, IEEE Transactions on Aerospace and Electronic Systems 20 (2) (1984) 183–198.
[75] A. Farina, E. Hanle, Position accuracy in netted monostatic and bistatic radar, IEEE Transactions on Aerospace and Electronic Systems 19 (4) (1983) 513–520.
[76] N.H. Nguyen, K. Dogancay, On the bias of pseudolinear estimators for time-of-arrival based localization, in: Proc. IEEE International Conference on Acoustics, Speech and Signal Processing (ICASSP), 2017, pp. 3301–3305.
[77] S. Nardone, A. Lindgren, K. Gong, Fundamental properties and performance of conventional bearings-only target motion analysis, IEEE Transactions on Automatic Control 29 (9) (1984) 775–787.
[78] V.J. Aidala, S.C. Nardone, Biased estimation properties of the pseudolinear tracking filter, IEEE Transactions on Aerospace and Electronic Systems 18 (4) (1982) 432–441.

[79] N.H. Nguyen, K. Dogancay, Closed-form algebraic solutions for 3-D Doppler-only source localization, IEEE Transactions on Wireless Communications 17 (10) (2018) 6822–6836.
[80] K. Dogancay, Bias compensation for the bearings-only pseudolinear target track estimator, IEEE Transactions on Signal Processing 54 (1) (2006) 59–68.
[81] K. Dogancay, 3D pseudolinear target motion analysis from angle measurements, IEEE Transactions on Signal Processing 63 (6) (2015) 1570–1580.
[82] K.C. Ho, Y.T. Chan, An asymptotically unbiased estimator for bearings-only and Doppler-bearing target motion analysis, IEEE Transactions on Signal Processing 54 (3) (2006) 809–822.
[83] N.H. Nguyen, K. Dogancay, Improved pseudolinear Kalman filter algorithms for bearings-only target tracking, IEEE Transactions on Signal Processing 65 (23) (2017) 6119–6134.
[84] N.H. Nguyen, K. Dogancay, Instrumental variable based Kalman filter algorithm for three-dimensional AOA target tracking, IEEE Signal Processing Letters 25 (10) (2018) 1605–1609.
[85] N.H. Nguyen, K. Dogancay, 3D AOA target tracking with two-step instrumental-variable Kalman filtering, in: Proc. IEEE International Conference on Acoustics, Speech and Signal Processing (ICASSP), Brighton, UK, 2019, pp. 4390–4394.
[86] K.C. Ho, W. Xu, An accurate algebraic solution for moving source location using TDOA and FDOA measurements, IEEE Transactions on Signal Processing 52 (9) (2004) 2453–2463.
[87] A.G. Lingren, K.F. Gong, Position and velocity estimation via bearing observations, IEEE Transactions on Aerospace and Electronic Systems 14 (4) (1978) 564–577.
[88] K. Dogancay, On the efficiency of a bearings-only instrumental variable estimator for target motion analysis, Signal Processing 85 (3) (2005) 481–490.
[89] N.H. Nguyen, K. Dogancay, L.M. Davis, Adaptive waveform and Cartesian estimate selection for multistatic target tracking, Signal Processing 111 (2015) 13–25.
[90] N.H. Nguyen, K. Dogancay, Single-platform passive emitter localization with bearing and Doppler-shift measurements using pseudolinear estimation techniques, Signal Processing 125 (2016) 336–348.
[91] C. Ma, R. Klukas, G. Lachapelle, An enhanced two-step least squared approach for TDOA/AOA wireless location, in: Proc. IEEE International Conference on Communications (ICC), Anchorage, AK, USA, vol. 2, 2003, pp. 987–991.
[92] N.H. Nguyen, K. Dogancay, Improved weighted instrumental variable estimator for Doppler-bearing source localization in heavy noise, in: Proc. IEEE International Conference on Acoustics, Speech and Signal Processing (ICASSP), Calgary, Canada, 2018, pp. 3529–3533.
[93] L. Ljung, T. Soderstrom, Theory and Practice of Recursive Identification, MIT Press, Cambridge, MA, 1983.
[94] S. Haykin, Adaptive Filter Theory, third ed., Prentice-Hall, Englewood Cliffs, NJ, 1996.
[95] J.A. Nelder, R. Mead, A simplex method for function minimization, The Computer Journal 7 (4) (1965) 308–313.
[96] F. Athley, Threshold region performance of maximum likelihood direction of arrival estimators, IEEE Transactions on Signal Processing 53 (4) (2005) 1359–1373.
[97] D. Koks, Passive geolocation for multiple receivers with no initial state estimate, Tech. Rep. DSTO RR-0222, Defence Science & Technology Organisation, Edinburgh, SA, Australia, 2001.

[98] K. Dogancay, Bearings-only target localization using total least squares, Signal Processing 85 (9) (2005) 1695–1710.
[99] M. Gavish, A.J. Weiss, Performance analysis of bearing-only target location algorithms, IEEE Transactions on Aerospace and Electronic Systems 28 (3) (1992) 817–828.
[100] A.S. Goldberger, Econometric Theory, Wiley, New York, NY, 1964.
[101] J.M. Mendel, Lessons in Estimation Theory for Signal Processing, Communications, and Control, Prentice-Hall, Englewood Cliffs, NJ, 1995.
[102] A. Weil, Number Theory: An Approach Through History From Hammurapi to Legendre, Birkhauser, Boston, MA, 2001.
[103] W. Feller, An Introduction to Probability Theory and Its Applications, vol. 1, third ed., Wiley, New York, NY, 1968.
[104] D. Musicki, R. Kaune, W. Koch, Mobile emitter geolocation and tracking using TDOA and FDOA measurements, IEEE Transactions on Signal Processing 58 (3) (2010) 1863–1874.
[105] M.L. Fowler, X. Hu, Signal models for TDOA/FDOA estimation, IEEE Transactions on Aerospace and Electronic Systems 44 (4) (2008) 1543–1550.
[106] A. Yeredor, E. Angel, Joint TDOA and FDOA estimation: a conditional bound and its use for optimally weighted localization, IEEE Transactions on Signal Processing 59 (4) (2011) 1612–1623.
[107] E. Weinstein, D. Kletter, Delay and Doppler estimation by time-space partition of the array data, IEEE Transactions on Acoustics, Speech, and Signal Processing 31 (6) (1983) 1523–1535.
[108] Y.T. Chan, K.C. Ho, A simple and efficient estimator for hyperbolic location, IEEE Transactions on Signal Processing 42 (8) (1994) 1905–1915.
[109] K. Dogancay, N.H. Nguyen, Low-complexity weighted pseudolinear estimator for TDOA localization with systematic error correction, in: Proc. European Signal Processing Conference (EUSIPCO), Budapest, Hungary, 2016, pp. 2086–2090.
[110] K. Dogancay, Emitter localization using clustering-based bearing association, IEEE Transactions on Aerospace and Electronic Systems 41 (2) (2005) 525–536.
[111] S.R. Drake, K. Dogancay, Geolocation by time difference of arrival using hyperbolic asymptotes, in: Proc. IEEE International Conference on Acoustics, Speech and Signal Processing (ICASSP), Montreal, Canada, vol. 2, 2004, pp. 361–364.
[112] Y. Huang, J. Benesty, G.W. Elko, R.M. Mersereati, Real-time passive source localization: a practical linear-correction least-squares approach, IEEE Transactions on Speech and Audio Processing 9 (8) (2001) 943–956.
[113] H.C. So, Y.T. Chan, F.K.W. Chan, Closed-form formulae for time-difference-of-arrival estimation, IEEE Transactions on Signal Processing 56 (6) (2008) 2614–2620.
[114] K.C. Ho, Bias reduction for an explicit solution of source localization using TDOA, IEEE Transactions on Signal Processing 60 (5) (2012) 2101–2114.
[115] R.J. Barton, D. Rao, Performance capabilities of long-range UWB-IR TDOA localization systems, EURASIP Journal on Advances in Signal Processing 2008 (1) (2007) 1–17.
[116] N.H. Nguyen, K. Dogancay, Computationally efficient IV-based bias reduction for closed-form TDOA localization, in: Proc. IEEE International Conference on Acoustics, Speech and Signal Processing (ICASSP), Calgary, Canada, 2018, pp. 3226–3230.

Index

FSC
www.fsc.org
MIX
Papier | Fördert
gute Waldnutzung
FSC® C083411

Kolibri 360 GmbH
Ferdinand-Jühlke-Straße 7
99095 Erfurt, Deutschland
produktsicherheit@kolibri360.de